Cytogenetics - Classical and Molecular Strategies for Analysing Heredity Material

Edited by Marcelo L. Larramendy and Sonia Soloneski

Published in London, United Kingdom

IntechOpen

Supporting open minds since 2005

Cytogenetics - Classical and Molecular Strategies for Analysing Heredity Material
http://dx.doi.org/10.5772/intechopen.92942
Edited by Marcelo L. Larramendy and Sonia Soloneski

Contributors
Vishnu Sharma, Tarun Kumar Kumawat, Garima Sharma, Manish Biyani, Rashi Garg, Sanjay Kumar, N. Abenthung Kithan, Asikho Kiso, Yahya Benbouchta, Khalid Sadki, Ahmed Afailal Tribak, Weihua Wang, Xiangli Niu, Yanping Lao, Yan Sun, Mounia Bendari, Nisrine Khoubila, Sofia Sraidi, Koluru Honnegowda Honnegowda Venkatesh, Roberto Ferreira Artoni, Daniele Aparecida Matoso, Maelin Da Silva, Hallana Cristina Menezes Da Silva, Eliana Feldberg, Maelin Silva, Vladimir Pavan Margarido, Tatiana Zybina

First published in London, United Kingdom, 2021 by IntechOpen
IntechOpen is the global imprint of INTECHOPEN LIMITED, registered in England and Wales, registration number: 11086078, 5 Princes Gate Court, London, SW7 2QJ, United Kingdom
Printed in Croatia

British Library Cataloguing-in-Publication Data
A catalogue record for this book is available from the British Library

Additional hard and PDF copies can be obtained from orders@intechopen.com

Cytogenetics - Classical and Molecular Strategies for Analysing Heredity Material
Edited by Marcelo L. Larramendy and Sonia Soloneski
p. cm.
Print ISBN 978-1-83968-941-3
Online ISBN 978-1-83968-942-0
eBook (PDF) ISBN 978-1-83968-943-7

For EU product safety concerns:
IN TECH d.o.o., Prolaz Marije Krucifikse Kozulić 3, 51000 Rijeka, Croatia, info@intechopen.com or visit our website at intechopen.com.

Meet the editors

Marcelo L. Larramendy, Ph.D., is a Professor of Molecular Cell Biology at the School of Natural Sciences and Museum, National University of La Plata (UNLP), Argentina. He is also a senior researcher of the National Scientific and Technological Research Council of Argentina. Dr. Larramendy is a former member of the Executive Committee of the Latin American Association of Environmental Mutagenesis, Teratogenesis and Carcinogenesis. He is the author of more than 450 contributions, including scientific publications, research communications, and conferences worldwide. He is a recipient of several national and international awards. Dr. Larramendy is a regular lecturer at the International A. Hollaender Courses organized by the International Association of Environmental Mutagenesis and Genomics (IAEMS) and a former guest scientist at the National Institutes of Health (NIH) and University of Helsinki, Finland. He is an expert in genetic toxicology and is, or has been, a referee for more than twenty international scientific journals. He was also a member of the International Panel of Experts at the International Agency for Research on Cancer (IARC, WHO, Lyon, France) in 2015 for the evaluation of DDT, 2,4-D and Lindane. Presently, Dr. Larramendy is head of the Laboratory of Molecular Cytogenetics and Genotoxicology at the UNLP.

Sonia Soloneski has a Ph.D. in Natural Sciences and is Assistant Professor of Molecular Cell Biology at the School of Natural Sciences and Museum of La Plata, National University of La Plata (UNLP), Argentina. She is a member of the National Scientific and Technological Research Council (CONICET) of Argentina in the genetic toxicology field, the Latin American Association of Environmental Mutagenesis, Teratogenesis and Carcinogenesis (ALAMCTA), the Argentinean Society of Toxicology (ATA), the Argentinean Society of Biology (SAB), and the Society of Environmental Toxicology and Chemistry (SETAC). She has authored more than 380 contributions in the field, including scientific publications in peer-reviewed journals and research communications. She has served as a review member for more than thirty scientific international journals. She has been a plenary speaker in scientific conferences and a member of scientific committees. She is a specialist in issues related to genetic toxicology, mutagenesis, and ecotoxicology.

Contents

Preface

As stated in our previous book *Cytogenetics - Past, Present and Further Perspectives*, published by IntechOpen in 2019, the classical definition of cytogenetics presented in Merriam-Webster, as employed since 1930, is "...a branch of biology that deals with the study of heredity and variation by the methods of both cytology and genetics." The medical definition of cytogenetics includes the study of chromosomes, which are the visible carriers of heredity material. Taken together, these definitions convert cytogenetics into a fusion science that joins cytology (i.e., the study of cells) with genetics (i.e., the study of inherited variation).

We should consider that more than seventy years have passed since that moment in 1952 when Tjio and Levan discovered the diploid number of chromosomes in humans to be 46. During this time, as can be expected, continuous advances in knowledge have resulted in the development of new techniques, ranging from initial conventional banding techniques to current molecular methodologies. Through combining these traditional and molecular techniques, cytogenetics has emerged as an essential tool in the diagnosis of various genetic disorders, leading the way for potential management and treatment.

The book presents recent advances and strategies in the field of cytogenetics, paying special attention to methodological achievements developed worldwide that have driven the field forward. The contributors clearly discuss several concepts and approaches useful for understanding chromosomal structure and function at its various levels, highlighting chromosomes as visible carriers of heredity material.

This book is organized into eleven chapters; it begins with a chapter presenting a detailed karyotyping investigation of three species of mulberry belonging to the genera *Morus* spp., including studies on basic chromosome number, ploidy levels, chromosomal associations and configurations, as well as meiotic behaviours of mulberry varieties. The second chapter presents a complete study analyzing the composition and nature of the heterochromatin in different species of electrical fish from the genera *Gymnotus*, which serves as a representative endemic organism of the Neotropical region. This information helps one to comprehend the genetic structures, evolution, and systematics of the richest and most diverse Neotropical fish groups. The third chapter provides an update regarding the relationship between constitutive heterochromatin and the inactivation of transposable elements, focusing on species of fish from the Amazonian region. The authors show that the species of Amazonian fish analyzed seem to respond dynamically, and with remarkable similarity, to a range of stressful stimuli. The fourth chapter describes conventional and molecular banding techniques routinely employed for detecting and quantifying chromosomal aberrations during DNA sample screening. The authors highlight a number of topics: the karyotype and its components, karyotype trends, evolution and its role in speciation, banding patterns and techniques, chromosome differentiation and linearization, banding applications, detection and analysis of chromosomal aberrations, types of chromosomal and chromatid aberrations and the mechanism of formation of these aberrations and breaks for karyotype evolutionary trends, the applications of these processes in both diagnostic and functional scenarios, and potential applications in the assessment of various risks faced by the genetic constitution of eukaryotic cells. The fifth chapter highlights the strengths and weaknesses of employing trophoblastic cells as an interesting model to investigate the role of gene copy number modifications and changes in gene expression that are important for normal versus abnormal cell differentiation. The sixth chapter describes the role of high-resolution next-generation sequencing techniques for detecting and quantifying chromosome abnormalities employed in preimplantation genetic testing. The authors highlight that one-step blastocyst biopsy can improve both blastocyst implantation and live birth rate in in-vitro fertilization patients, suggesting the one-step biopsy method is superior to the two-step method. Finally, the authors remark that the screening of these chromosome abnormalities may reduce embryo implantation failure, early miscarriage, birth defects, developmental delay and intellectual disability. The seventh chapter describes commonly employed methodologies for elucidating chromosomal aberrations, the prognostic impact of recurrent chromosomal abnormalities, and risk stratification in haematological disease genetics. Additionally, this chapter presents various types of chromosomal abnormalities identified in many haematological diseases, such as acute myeloid leukaemia, acute lymphoid leukaemia, myelodysplastic syndrome, multiple myeloma, and myeloproliferative diseases, thus highlighting their role in clinical phenotyping and prognosis. The eighth chapter discusses the employment of conventional cytogenetic and molecular analyses involved in the pathogenesis of acute lymphoblastic leukaemia, one of the most common childhood cancers. The chapter highlights how conventional cytogenetic techniques have undergone significant advancements, emerging into molecular cytogenetic technologies, such as fluorescence in situ hybridization, multiplex ligation-dependent probe amplification, array comparative genomic hybridization, and next-generation sequencing, all of which serve as innovative techniques for diagnosis, risk stratification, disease monitoring, and treatment selection. The ninth chapter reviews advances in cytogenetics and genetics in terms of myelodysplastic syndromes, focusing on the role of the conventional karyotype and new genomic technologies, such as single-nucleotide polymorphism array and next-generation sequencing involved in myelodysplastic syndromes, to discover new genetic alterations and develop targeted therapies. The tenth chapter highlights the importance of the identification of the Philadelphia chromosome in chronic myeloproliferative syndromes in terms of diagnosis and prognosis to improve the quality of healthcare provided to patients. Finally, the book concludes with a chapter reviewing how environmental changes in the frontier of mutations or misregulations of cytokeratin genes are involved in the pathogenesis of innate keratinizing disorders, such as cutaneous tissue fragility, skin hypertrophy, and malignant transformation.

The editors of *Cytogenetics - Classical and Molecular Strategies for Analysing Heredity Material* are enormously grateful to all the contributing scientists for sharing their knowledge and insights in this interdisciplinary book project; they have made extensive efforts in arranging the information of each chapter. We hope that the information presented in this book will meet the expectations and needs of all those interested in the various aspects of cytogenetics, including scientists, physicians, pharmacologists, and students, among others. We especially hope that this book will guide those in the field to make new discoveries through employing novel and prior methods of investigation in their future research using an understanding of both basic and applied aspects of cytogenetics.

Sonia Soloneski Ph.D. and Marcelo L. Larramendy Ph.D.
School of Natural Sciences and Museum,
National University of La Plata,
La Plata, Argentina

Chapter 1

Studies on Basic Chromosome Number, Ploidy Level, Chromosomal Association and Configuration and Meiotic Behavior in Mulberry (*Morus* Spp.)

Koluru Honnegowda Venkatesh

Abstract

Mulberry leaves are primary food for silkworm, *Bombyx mori* L. to feed silkworms and harvest quality silk cocoons. Mulberry belongs to family Moraceae and includes 60 species found distributed in both Hemisphere. In mulberry, chromosome numbers are varies from 2n = 28 to 22n = 308 (Diploid to Decosoploid) with ploidy level x to 22x. Based on chromosome numbers and meiotic behaviors x = 14 has been considered as basic chromosome numbers of the genus. In the present study, two diploids, two uneuploids, two triploids and two teteraploids mulberry varieties were selected for detailed chromosomal numbers and meiotic behaviors belongs to three species, namely *Morus indica, Morus alba* and *Morus latifolia.* Varieties, Vishaala and Kosen were diploids with 2n = 2x = 28 chromosomes and varieties Ber-S_1 and S_{13} were uneuploids with 2n = 30 chromosomes belongs *Morus indica.* Varieties NAO Khurkul and KPG-1 were triploids with 2n = 3x = 42 chromosomes belongs to *Moru alba* and varieties Kokuso and Icheihei were tetraploids with 2n = 4x = 56 chromosomes. Diploids and uneuploids were showed normal meiosis with high pollen fertility and triploids and teteraploids were showed abnormal meiosis with low pollen fertility, due to virtue of higher ploidy level have been discussed in this chapter.

Keywords: Mulberry, chromosomes, polyploids meiotic behavior

1. Introduction

Sericulture is as an important agriculture-based, labor intensive, export-oriented cottage industry, introduced more than 200 years ago in India. This industry consists of several sectors or processes that are linked to one another like a chain. They are mulberry cultivation, silkworms egg production, silkworm rearing, harvesting of cocoons, silk reeling, twisting and weaving and manufacturing silk fabrics. Mulberry silk is produced from silkworms (*Bombyx mori* L.), which form the base of silk production. Mulberry is a fast growing plant and hence farmers can harvest 5–6 silkworm crops in a year at an interval of 26–28 days. The marginal and small farmers opted for sericulture since it was a remunerative crop as compared to other competing crops like Raagi, Jowar, Paddy, Potato and other vegetable crops.

Morus L. is an important genus of the family Moraceae under the order Urticales [1, 2] established the genus *Morus* with seven species viz., *Morus. alba, M. nigra, M. rubra, M. indica. M. tartarica, M. papyrifera* and *M. tinctoria.* Later, a number of species have been discovered by various workers from different parts of the world. Mulberry is a dicot, mesophytic, heterozygous and cross pollinated plant.

Importance of cytogenetical studies is very well understood in all most all agricultural crops. Even in mulberry breeding emphasis has been laid to understand the cytology of genotypes used as parents in breeding, in all sericultural advanced countries. In various mulberry genotypes, basic gametic and somatic chromosome numbers suggesting the ploidy level [3, 4]. These Cytogenetical data are useful to mulberry breeders in identifying and evolving promising genotypes and selecting the suitable species/varieties for commercial exploitation. Keeping this in view the identifying proper representation of genotypes of three species of mulberry has been discussed in this chapter.

2. Chromosome numbers

It is well established fact that cytological features are employed in differentiating and tracing the phylogeny of organisms. Cytotaxonomy based on chromosomal characteristics was most popular in plant taxonomy between 1930 and 1960 [5]. In such taxonomic studies chromosome number, chromosome morphology, chromosomal association, chromosome behavior and cytochemistry, etc., have yielded valuable results and revolutionized the phylogenetic interpretations.

Cytological studies in mulberry have been carried out in all sericulturally advanced countries. In view of their importance in breeding programs, chromosome numbers, chromosomal association and meiotic behaviors of various diploid, triploid, tetraploid and uneuploid genotypes of mulberry have been studied.

2.1 Polyploidy

The organisms having more than two genomes or two sets of chromosome in their somatic cells are called as polyploids. Among plants and animals, the polyploidy occurs in a multiple series of 3, 4, 5, 6, 7, 8, etc., of the basic chromosome numbers or genomes number and thus it causing triploidy, tetraploidy, pentaploidy, hexaploidy, heptaploidy, octaploidy, respectively. Polyploidy is most common among angiosperms.

The phenomenon of polyploidy is one of the widespread and distinctive features, which has played a major role in the evolution of higher plants. It plays important role in the natural selection and better adaptability of species in the new ecological niches.

In mulberry, polyploidy breeding techniques are found to be more suitable than mutation breeding techniques. A number of varieties have been developed in sericulturally advanced countries like Japan [6], China [7] and also in India [8]. In India, triploid mulberry varieties like TR_8 and TR_{10} have been recommended in hilly areas of eastern states.

Polyploidy may arise by several ways. (1) The egg may be fertilized by more than one sperm. If normal haploid egg is fertilized by two haploid sperms a triploid will result. (2) There may be failure of mitosis. (3) Triploids may arise as a result of fertilization of unreduced gametes. Diploid gametes arise because of failure in meiosis. If these gametes are fertilized by haploid sperms, triploids are formed.

(4) An autotetraploids may arise either by the doubling of chromosomes or by fertilization between two diploid gametes.

2.2 *Morus indica*

2.2.1 Variety Vishala

Vishala is developed by Central Sericultural Research Institute in Mysuru. It is fast growing variety, under ideal agro-climatic conditions (**Figure 1**). This variety yields 34,000 to 60,000 kgs and 14, 000 to 20,000 kgs of leaves/hector/year under irrigated and rain fed conditions respectively. Leaves are larger, dark green in color, unlobed and retain high moisture content. Stomatal frequency was found to be 260. 51/mm^2 & size 138.30μm^2.

2.2.2 Variety Kosen

This variety is evolved from cross pollinated hybrids. It is medium branching and fast growing in tropical conditions and it is good rooter (**Figure 2**). It possesses wide acclimatization in different agro-climatic conditions. This variety yields 26,000 to 40,000 kgs and 13, 000 to 18,000 kgs of leaves /hector/year under irrigated and rain fed conditions respectively. Leaves are larger, dark green in color, unlobed and retain high moisture content. Stomatal frequency was found to be 234. 60/mm^2 & size 126.40μm^2.

Each species of plants and animals is characterized by a particular chromosome complement or a set of genome, represented once in gametic haploid cell and twice in somatic diploid cells.

Genus *morus* exhibits a high degree of polyploidy ranging from diploid (2n = 2x = 28) to Decosoploid (2n = 22x = 308). Accordingly, its various species show chromosome numbers ranging 28 to 308. Majority of the species are diploids and polyploidy complexes were reported in many species.

Figure 1.
Variety Vishala.

Figure 2.
Variety Kosen.

Varieties, Vishala and S_{13} are belonging to Indian species and many cultivars belong to this species. Both the varieties were found to be 28 chromosomes ($2n = 2x = 28$) in their shoot somatic cells, during mitotic division, thereby confirming their diploid status. The diploid nature is related with fertility, normal growth, great vigorosity, adoptability and survivality of the diploid species. The typical characters of diploid are good elongation of branches and roots, good root initiation ability, good regenerating ability of buds, high yielding potential and easiness of raising saplings. Feeding value of leaves is highest for diploid, followed by triploid and tetraploid.

2.2.3 Variety Ber-S_1

It is an evolved variety from Berhampore Institute and showed uneuploid nature with $2n = 30$ chromosomes (**Figure 3**). Morphologically uneuploid varieties are almost similar to the diploids in all parameters except minor variations. The uneuploid nature is related with fertility, normal growth, great vigorosity, adoptability and survivality of the uneuploids varieties. It possesses wide acclimatization in different agro-climatic conditions. This variety yield 24, 000 to 38, 000 kgs and 11, 000 to 16, 000 kgs of leaf/hector/year under and rainfed conditions respectively. Leaves are larger, dark green, unlobed and retain high moisture content. Stomatal frequency was found to be 238.20/mm^2 & size 116.00μm^2.

2.2.4 Variety S_{13}

It is selected from open pollinated hybrid (OPH) of $Kanva_2$ during 1986. This variety is characterized by short internodes and having a capacity of produces large numbers of branches. Leaves are thick and dark green unlobed with smooth surface. This variety best suited for rainfed condition and yield 16,000 to 18,000 kgs of leaf /hector/year (**Figure 4**). Leaves of diploids and uneuploids varieties are found to be

Figure 3.
Variety Ber-S_1.

Figure 4.
Variety S_{13}.

succulent, rich in moisture and nutrient contents when compared to triploids and tetraploid varieties and are suitable to silkworm larvae [9]. Stomatal frequency was found to be 210.00/mm^2 & size 128.00μm^2.

Varieties, Ber-S_1 and S_{13} are belonging to Indian species and many cultivars belong to this species. Both the varieties were found to be 30 chromosomes

(2n = 30) in their shoot somatic cells, during mitotic division, thereby confirming their uneuploid nature. Uneuploids plants have incomplete genomes. Individual chromosomes may either be less than diploid number (monosomic and nullisomic), or more than the diploid number (polysomic). Uneuploid chromosome number recorded in present work as well as reported by others are mainly due to extensive vegetative propagation followed for the multiplication of *Morus* spp. Therefore, the genus *Morus* has monobasic number x = 14. The polyploid numbers found in this taxa must have derived from this base number (x = 14) an account of auto and allopolyploidizaton. No doubt, vegetative propagation has helped for the perpetuation of uneuploids rather than their origin. Out breeding has played an important role in the origin of uneuploids due to the formation of gametes with unbalanced chromosome numbers.

2.3 *Morus alba*

2.3.1 Varieties: NAO Khurkul and KPG_1

These are exotic varieties grown in both tropical and temperate conditions. It possesses wide acclimatization in different agro-climatic conditions (**Figures 5** and **6**). These varieties yields 30,000 to 35,000 kgs of leaves/hector/year under rain fed land. Leaves are medium, unlobed, light green in color and high moisture retention capacity. Stomatal frequency was found to be $290.80/mm^2$ & size $190.80 \mu m^2$ and $262.44/mm^2$ & size $198.30\ \mu m^2$ respectively.

Both the varieties were found to be 42 chromosomes (2n = 3x = 42) in their shoot somatic cells, during mitotic division, thereby confirming their triploid status. Triploids produced from the cross between diploids and tetraploids. Triploid is the characterized by three genomes or three sets of chromosomes in each somatic cells. Thus, a triploid originates by the fusion of haploid gamete (n) with a diploid gamete (2n), the later of which may be originated by irregularities during meiosis. Most of the naturally occurring polyploids are either allo-polyploids or complexes

Figure 5.
Variety NAO Khurkul.

Figure 6.
Variety KPG$_1$.

between allopolyploids and auto-polyploids. Agronomic ally, the triploid mulberry varieties are known to be superior to other ploidy levels in rooting, chemical components and silkworm rearing performance. They also posses resistance to cold and high temperature.

2.4 *Morus latifolia*

2.4.1 Varieties Kokuso and Icheihei

Morphological characters of these varieties are entirely different from other varieties (**Figures 7** and **8**). *Morus latifolia is e*asily distinguishable from *M. alba* and *M. indica* by cylindrical fruits and hairy leaves. Leaves of tetraploid varieties are smaller, thin, rough, lobed, highly dissected, less moisture content and coriaceous due to the presence of many cystoliths, on the leaf surface, the calcium oxalate deposit and have a very detrimental effect on the palatability to silkworm larvae. Hence, these leaves are considered as inferior quality and are not suitable to silkworm larvae. The leaves' upper surface is dark green and lustrous with a pale green under surface. Stomatal frequency was found to be 310.22/mm^2 & size 238.10μm^2 and 340.40/mm^2 & size 260.30μm^2 respectively.

The above description suggests that diploid and triploid plants of this genus may exist anywhere in the world and that the morphological characteristics of diploid and triploid plants will be different to a certain extent from those of the polyploid ones.

The number of chromosomes of both the varieties at metaphase of somatic division were 2n = 4x = 56, which means both are teteraploids. Tetraploid may be originated by the somatic doubling of the chromosome numbers or by union of unreduced gametes.

Figure 7.
Variety Kokuso.

Figure 8.
Variety Icheichei.

3. Micro-morphological features

Stomata are the important anatomical and physiological characteristics of a leaf which control the rate of CO_2 movement from air to the chlorophyllus tissue within the leaf and thereby influence the rate of CO_2 exchange [10]. Stomatal size and frequency are important parameters in selecting drought resistant genotypes and these are also believed to regulate the leaf yield. As such, the Stomatal frequency has been studied in these taxa to correlate with cytological information recorded in the present investigation. In general, Stomatal frequency and size was found to be high in tetraploid when compared to triploid and diploid varieties.

Stomatal frequency was correlated with drought resistance and disease resistance [11]. Further, the Stomatal frequency was found to be related to the dry weight of the aerial parts of the plants and diffusive resistance [12]. Smaller stomata with lesser frequency per unit area make mulberry leaves become more succulency, palatable and are suitable for silkworms. [13] reported that, moisture retention capacity will be higher in those mulberry varieties possessing smaller stomata and lower stomatal frequency. Similarly we found average more number of chloroplasts in the guard cells were found be 10–12 in diploids and uneuploids compared to the triploids and tetraploids 6–8. In diploids and uneuploids number chloroplast were almost double than triploids and uneuploid.

As mulberry is a heterogeneous, predominantly unisexual, cross-pollinated and as well as vegetatively propagated perennial plant, it exhibits a wide range of variations in each and every phenotypic characteristic. Often the same genotype reveals the presence of both and unlobed leaves.

The genotypic differences in trichome density observed in the present study provide evidence to such observations in the field. The dense cover of trichomes in shoot tip and leaves of diploids may pose resistance to thrips attack in diploid and uneuploid mulberry varieties. On the other hand shoot tip and leaves of triploid and tetraploid mulberry varieties are less protected with sparse trichome density and hence may become susceptible to the thrips attack. It is also reported that diploids has high drought tolerance capacity than triploids [14]. So the dense trichomes may provide a better barrier to excessive water loss than sparse trichomes.

Trichomes are epidermal structures occur on stem, petioles & leaves. Diploid and uneuploid varieties showed highest trichomes density in both stem and leaves. On the other hand triploid and tetraploid varieties showed sparsh trichome density in both stem and leaves. Similar to various other anatomical feature cystolith frequency was lower in diploids and uneuploid compared to triploid and tetraploids. Triploids and Tetraploids showed distinctly high frequency and large sized cystolith, similar to stomatal frequency and size. In the evaluation of mulberry varieties the frequency and size of the idioblast has been taken as a parameter instead of frequency and size of cystolith [15].

3.1 Chromosomal association and configuration

Varied chromosomal associations and configurations recorded in the present study indicate closer homology between associated chromosomes. Depending on the ploidy level and pairing of chromosomes bi-, tri- and quadrivalents showed in different cofigurational shape. In both diploid and uneuploid varieties the bivalents frequency was high and the regular univalent's appears to be a mere matter of chance. Most of the cultivated diploid mulberry varieties showed 14 bivalents at diakinesis and metaphase-I. Theoretically diploids should forms more bivalents in meiosis due to presence of two homologous chromosomes. Bivalents and univalents were observed ' - - ' or "⁀‿" and " -" types.

In triploid varieties the trivalent frequency was high and low frequency of bi and univalent's and no higher chromosome configuration was found. On the contrary some univalent's and bivalents were observed.

The alignment of centromeres of trivalents and univalents appears to be random and hence the assortment of chromosomes was also random. Hence majority of the cells showed unequal separation of chromosome at anaphase-I. Trivalents were observed "- - -" or ">-' types.

In tetraploid varieties various types of chromosomal associations ranging from uni- to quadrivalents are observed during diakinesis and metaphase-I. Frequencies of quadrivalents are more than that of uni-, bi- and trivalents. However, the occurrence of multivalent such as tri- and quadrivalents indicates its polyploidy nature. Quadrivalents were observed "⁀⁀/‿‿" or "- - - -" types.

Among different chromosomal associations, the bivalents in diploid and uneuploids, trivalents in triploid and quadrivalents in tetraploid were frequent. Majorities of quadrivalents were of ring type and other were chain types.

The concept of assessing the nature of polyploidy based on the sole criterion of multivalent frequency appears to have a limited value. Several workers have shown that, pairing of chromosomes is mainly governed by genes.

4. Meiotic behavior

4.1 *Morus indica* variety Vishaala

Meiotic behaviors in the present study revealed that microsporogenesis was normal in diploid varieties. These varieties showed regular pairing of chromosomes in majority of PMCs similar to many other diploid mulberry varieties [16] and 14 bivalents. The present investigation also showed there were no secondary association of chromosomes and multivalents in diakinesis and metaphase- I. The behavior of chromosome in metaphase is an index of fertility. Occurrence of normal bivalents results in the regular separation at anaphase and regular meiosis leading to high percentage of pollen fertility.

4.2 *Morus indica* variety S_{13}

Uneuploid varieties have incomplete genomes. Individual chromosome may either be less than the diploid number (*monosomics* and *nullisomic*) or more than the diploid number (polysomic). These varieties revealed marginal irregular meiosis and pollen fertility was also slightly less in these varieties.

4.3 *Morus alba* varieties NAO Khurkul and KPG_1

Triploid varieties are characterized by highly irregular meiosis and very low pollen fertility. Irregular meiosis is the almost common feature of triploids in all plants. These varieties showed high frequency of trivalents which suggests the fair homology between its constituent genome and autotriploid nature of these varieties [17, 18]. Such irregular meiosis has been reported in other natural triploid mulberry varieties also [19, 20]. The most common aberrations observed includes the occurrence of univalent's, laggards, stickiness, precocious movement and in anaphase unequal number of chromosome segregate to the poles which again attribute to irregularities in chromosome pairing and their alignment on equatorial plate. These irregular meiosis leads to formation of aberrant, unbalanced microspore and finally resulted in the reduction of pollen fertility. Formation of such aberrant sporads in some mulberry triploids has been recorded by [21].

4.4 *Morus latifolia*

4.4.1 Varieties Kokuso and Icheihei

Meiotic behavior was irregular in these varieties. Theoretically tetraploid forms revealed more quadrivalents in meiosis due to the presence of four homologous chromosomes. Occurrence of high frequency of tetravalent in both the varieties indicates their relatively more stable autotetraploids nature. The presence of low frequency of trivalents and univalent's along with bivalents also indicates segmental homology of chromosomes and the allotetraploid nature of these varieties have been observed by [22]. Both the varieties showed low pollen fertility. The reduced pollen fertility of tetraploids could be attributed to association of chromosomes into multivalent during synopsis and other meiotic abnormalities which invariably results in loss of chromatin material [23–25]. Chromosome segregates unequally during anaphase leading to imbalanced chromosome complement in microspores and also leads to pollen sterility.

During the study of meiosis, nature of pollen mother cells (PMCs) at pre-meiotic interphase, nature of chromosomes at early stages of prophase- I, pairing behavior of chromosomes at diakinesis and metaphase-I, segregation of chromosomes at anaphase-I, separation of chromatids at anaphase-II, type and variations in tetrad, pollen size and pollen stainability were studied. The frequency of different types of chromosomal associations scored in 50 meiotic cells at metaphase. The bivalents showed 1 or 2 chaisma could not be distinguished, as the chromosomes were shorter.

Basic chromosome numbers, ploidy level, chromosomal association and configurations and meiotic behaviors of eight mulberry varieties were studied. The present research work revealed the diploid chromosome number of $2n = 2x = 28$ (Vishaala & Kosen), uneuploid chromosome number of $2n = 30$ (Ber-S_1 & S_{13}), triploid chromosome number of $2n = 3x = 42$ (NAO Khurkul & KPG_1) and tetraploid chromosome number of $2n = 4x = 56$ (Kokuso & Icheihei). High percentage of pollen fertility in diploid plants with $2n = 28$ chromosomes indicate the dibasic nature of these taxa.

Variation in the chromosome pairing, assortment and pollen stainability in different varieties of mulberry indicate that they vary in genetic status. The basic number $x = 14$ appears to deep sited in the genus *Morus* and other numbers might have derived through secondary polyploidy, hybridization followed doubling of chromosome and uneuploid alteration.

5. Conclusion

The present investigation was carried out with objectives of generating useful information on Studies on basic chromosome number, ploidy level, chromosomal association and configuration and meiotic behaviors of eight mulberry varieties (Genotypes).

The mulberry varieties studied (Genotypes) have unraveled diversity with respect of branching pattern, leaf yield, leaf color, size, lobation, shape, stomatal frequency and size, chloroplasts in the guard cells and trichomes density. The taxa studied were taxonomically distinct. The interaction between genotype and the environment is largely responsible for the diversity observed.

Based on the present findings diploids have $2n = 28$, uneuploids have $2n = 30$, triploids have $3n = 42$ and teteraploids have $4n = 56$ chromosomes. Meiosis has been found to be regular with high frequency of bivalents, high pollen stainability and development normal pollen grain in diploids compare to irregular meiosis characterized by the presence of high frequency of trivalents and univalent's, low pollen stainability and aberrant pollen formation in triploids and teteraploids. The reduced pollen fertility has been attributed to various meiotic anomalies. The role of chromosome repatterning coupled with polyploidy (both euploids and uneuploids) structural alternation of chromosomes and mutations in the evolution of the genus *Morus* is suggested.

It may be concluded that more intensive biosystematics studies involving a large numbers of varieties /genotypes of a species definitely throw more light on the phylogeny and systematics of mulberry species. Such information's are highly essential and useful in mulberry improvement programmes.

Author details

Koluru Honnegowda Venkatesh
Plant Breeding and Genetics Laboratory, Department of Life Sciences/Biological Sciences, Jnanabharathi Campus, Bangalore University, Bengaluru, Karnataka, India

Address all correspondence to: khvenki1972@gmail.com

References

[1] Airy Shaw. A dictionary of flowering plants and ferns. 8th edition (J. C. Willis). Cambridge University Press, London, 1993

[2] Linneaus C. Species plantorum, Stockholm, 1975; 986.

[3] Dandin SB, Basavaiah, Mala V Rajan, Suryanarayana, (1989). Basic chromosome of the genus *Morus* L. A critical reappraisal. Pro. Conf. on Cytol. and Genet.,1989; **2**: 203-211

[4] Venkatesh KH. Cytogenetical investigation in the genus *Morus* L. Ph. D. Thesis, 2007; Bangalore University Bengaluru.

[5] Wanger, WH Jr., Plant taxonomy and modern systematics. Bioscience.,1968; **12**:96-100

[6] Tojyo I. Studies on the polyploid mulberry tree (IV) on the flower and pollen grains of one race of the *Morus nigra* L. *J. Seric.* Sci. Jpn.,1966; **35**(5): 360-364

[7] Yang JH and Yang XH. Breeding of artificial triploids in mulberry. Acta. Sericologia Sinica, 1989; **15**(2): 65-70

[8] Vijayan K, Chakraborti SP, Roy BN, Setua M, Qadri SMH. Cytomorphological characteristics of a cochicine induced chimeric mutant in mulberry. Cytologia, 1998; **63**:27-31

[9] Dzhafurov NA, Alekprov OR. Fodder qualities of the leaves of inbreed and interspecific polyploid mulberry hybrids, Shelk, 1978; **4**:5 (Ru)

[10] Wallace DK, Ozaban JL, Munger HM. Physiological genetics of crop yield. Ad. Agro., 1972; **24**:97-146

[11] Nautiyal S, Badola HK, Pal H, Negi DS. Plant responses to water stress. Changes in growth dry matter production, stomata frequency and leaf anatomy, Biol. Plant., 1994; **36**(1): 91-97.

[12] Goek Y, Dun GM. Stomatal length, frequency and distribution in *Bromus inermis* leaf. Crop Science., 1975; **15**(3) 283-286

[13] Basavaiah, Murthy TCS. Relation between anatomical features, moister content and moister retention capacity of leaf in some mulberry (*Morus* spp.) genotypes of different ploidy. Natl. Semi. Mulb. Seri. Res. November, 2001; 26-28, KSSR& DI, Bangalore, India:98

[14] Susheelamma BN, Jolly MS. Evaluation of morphological parameters associated with drought resistance in mulberry. Indian J. Seric., 1986; **25**(1):6-14

[15] Thangavelu K, Mukherjee P, Tikader A, Ravindran S, Goel AK, Ananda Rao A, Girish Naik V, Shekar S. Catalogue on mulberry (*Morus spp.)* germplasm, CSGRC, Hosur, Tamil Nadu, 1997; **1**: 1-236

[16] Dandin SB, Basavaiah, Mala V Rajan, Suryanarayana. Basic chromosome number of the Genus *morus* L. A critical reappraisal. Pro. Conf. on Cytol. and Genet. 1989; **2**:203-211

[17] Seki. Cytological studies in mulberry *Morus* I. Polyploidy of the mulberry trees with special reference to spontaneous occurrence of triploid plants. J. Fac. Text and Seri, Shinshu Univ., Ueda, Japan, 1959; **20**: 1-59

[18] Das BC, Verma RC, Sarkar A. Chromosome associations in natural and induced triploids of mulberry. Cell and Chrom. Res. 1984; 7(2): 60-61

[19] Basavaiah, Mala V Rajan, Dandin SB, Suryanarayana N, Sengupta. Chromosomal association and meiotic

behavior of four triploid varieties of mulberry (*Morus* spp.) *Cytologia*, 1990; **55**: 327-333

[20] Venkatesh KH. Cytogenetical investigation in the Genus morus L., Ph. D. Thesis 2007; Bangalore University, Bengaluru.

[21] [21]Osawa I. Cytological and experimental studies in *Morus* with special reference to triploid mutants. Bull. Imp. Sercult. Sta. Japan, 1920; **1**:318-366.

[22] Dwivedi NK, Sikdar AK, Dandin SB, Sastry CR, Jolly MS. (1986). Induced tetraploidy in mulberry. Morphological, anatomical and cytological Investigation in cultivar RFS_{135}, Cytologia, 1986; **51**:393-401

[23] Darlington CD. (1965). *Recent advances in Cytology.* J and A Chrucill Ltd. London, 1965; 85-134

[24] Gottschalk W. Open problems in polyploidy research. Nucleus, 1978; **21**:91-112

[25] Das BC, Prasad DN, Krishnaswamy S. (1970). Studies on anthesis in mulberry *Indian* J. Seric.,1970; **9**: 59-64

Chapter 2

Composition and Nature of Heterochromatin in the Electrical Fish (Knifefishes) *Gymnotus* (Gymnotiformes: Gymnotidae)

Maelin da Silva, Daniele Aparecida Matoso, Vladimir Pavan Margarido, Eliana Feldberg and Roberto Ferreira Artoni

Abstract

Fishes of the genus *Gymnotus* have been suggested as a good model for biogeographic studies in the South American continent. In relation to heterochromatin, species of this genus have blocks preferably distributed in the centromeric region. The content of these regions has been shown to be variable, with description of transposable elements, pseudogenes of 5S rDNA and satellite sequences. In *G. carapo* Clade, although geographically separated, species with 2n = 54 chromosomes share the distribution of many 5S rDNA sites, a unique case within the genus. Here, repetitive DNA sequences from *G. sylvius* (2n = 40) and *G. paraguensis* (2n = 54) were isolated and mapped to understand their constitution. The chromosome mapping by FISH showed an exclusive association in the centromeres of all chromosomes. However, the cross-FISH did not show positive signs of interspecific hybridization, indicating high levels of heterochromatic sequence specificity. In addition, COI-1 sequences were analyzed in some species of *Gymnotus*, which revealed a close relationship between species of clade 2n = 54, which have multiple 5S rDNA sites. Possibly, the insertion of retroelements or pseudogenization and dispersion of this sequence occurred before the geographic dispersion of the ancestor of this clade from the Amazon region to the hydrographic systems of Paraná-Paraguay, a synapomorphy for the group.

Keywords: FISH, Biogeography, Satellite DNA, rDNA 5S, C_0t-*1*

1. Introduction

Repetitive DNA sequences are broadly distributed in eukaryotes genomes [1] and are classified into two categories: 1) repetitive sequences arranged in tandem as satellite, minisatellite, or microsatellite DNAs composed of hundreds of base pairs repeated a thousand times or more in each genome; and 2) moderately to highly repetitive sequences spread throughout the genome as retroelements or transposable mobile elements [2].

Copies of repetitive sequences are commonly associated with heterochromatin regions that can be visualized by C banding. These sequences are extremely important to the functional and structural organization of the eukaryote genome, composing, for example, the pericentromeric heterochromatin regions [3, 4]. The heterochromatin in fish chromosomes is largely located in pericentromeric regions and has structural functions [5–7].

The studies about the location of repetitive sequences on chromosomes has broadened the knowledge of the structural organization of chromatin in fish, revealing an association of ribosomal DNA, telomeric sequences, transposition elements and satellite sequences in chromosomal rearrangements and weak break points [8, 9], in the fixation of sex chromosomes [10, 11], in the expansion of heterochromatin [12, 13] and in gene regulation [14]. This advances in molecular cytogenetics have demonstrated that repetitive DNA sequences are useful as chromosomal markers in studies of species evolution and can provided valuable information about sex chromosome systems and chromosomal rearrangements [15]. The mapping of Non-long terminal repeat (non-LTR) retrotransposable elements, the Rex in the fish species, for example, demonstrated strong FISH signals in heterochromatin regions [16].

Neotropical electric fish species, order Gymnotiformes, have shown their heterochromatin to be preferentially distributed in the pericentromeric regions of their chromosomes [13, 17, 18].

The investigation of repetitive sequences in *Gymnotus* seems promising for understanding the chromosomal evolutionary dynamics in the genus. DNA probes and chromosomal painting were applied to investigate chromosomal rearrangements in two species of the *G. carapo* complex, and rearrangements were found between the two species, involving several pairs of chromosomes, corroborating the existence of cryptic species in this group, in addition to the recent speciation between them [19, 20]. Analysis of the satellitome of some species by Next Generation Sequencing (NGS) revealed sets of conserved satellite sequences, but the CA, GA and GAG motifs when mapped revealed a useful band for identification of homologous chromosomes [21].

The ribosomal DNA mapping can also provide new answers about chromosomal evolution in the genus, and even serve as a tool for understanding geographic distribution patterns [22]. In the species that had the 18S rDNA mapped, the proposal of only one pair carrying the conserved nucleolus organizing regions (NORs) is plausible [17, 18, 23, 24]. Regarding the 5S rDNA, the group behaves as an attractive model for evolutionary studies, showing a species-specific pattern. The evolution of this gene family receives special attention in the species that comprise the group *G. carapo* with a diploid number of 54 chromosomes, which have from 14 to 19 pairs identified with this sequence [21]. This situation is totally different from the pattern observed for other species within the group and shared only by species carrying 2n = 54 chromosomes.

2. Biogeography of electric fish and repetitive DNA sequences

The complex history of the formation of the South American rivers is fundamental to explain the diversity and distribution of aquatic biota in this region. Successive continental geomorphological changes, such as the one that resulted in the formation of the Andes, associated with historic and biological factors allow the identification of patterns that led to the formation of the largest and most diverse freshwater ichthyofauna on the planet [25, 26]. Such changes alter the drainage scenario forming lakes, capture headwaters and basins of varying sizes, or even isolate populations for certain periods, favoring the diversification of biota by vicariance and allopatric, in addition to promoting subsequent drainage coalescence, leading to enrichment and contact of organisms [26, 27].

The Gymnotiformes order comprises electric fish or knifefishes. Endemic to the Neotropical region [28], which are widely distributed, from the Pampas in Argentina to Chiapas, Mexico and reach their greatest diversity and abundance in the Amazon basin [29–31]. Members of the Gymnotiformes order, are unique in their ability to produce and recognize electrical signals, never left the South America plate, since their electrosensory system is not capable of functioning in brackish or salt water [21].

Regarding the karyotype, *Gymnotus* is also a diverse group, with a diploid number ranging from 34 chromosomes, verified in *G. capanema* [33] to 54, found in several species of the *G. carapo* group [18, 21, 23–34]. Sexual chromosomal heteromorphism has already been recorded in some, and all records follow a turnover model, with little heterochromatin and of recent origin [24, 35].

Among the *Gymnotus*, an intriguing fact has been reported for species that have 2n = 54 chromosomes, present in the *carapo* Clade: *G.* cf. *carapo*, *G. inequilabiatus*, *G. paraguensis* and *G. mamiraua*. These species share the diploid number and also a large number of 5S rDNA sites [13], but they are geographically separated in different hydrographic basins. Though, *G.* cf. *carapo*, *G. inequilabiatus*, *G. paraguensis* occur in the Paraná-Paraguay basin, and *G. mamiraua* is distributed in the Amazon basin (**Figure 1**). The number of 5S DNAr sites in *G. paraguensis*, *G. inaequilabiatus* and *G.* cf. *carapo* are 19, 17 and 15 bearing pairs, respectively [36–38], and 14 pairs in *G. mamiraua* [22].

For *G. paraguensis* and *G. mamiraua*, the 5S rDNA are present in the pericentromeric region and when sequenced, a transposition element similar to Tc1 was identified in the NTS (non-transcribed spacer), suggested as one of the mechanisms used for the transposition of this rDNA in chromosomes [22, 36]. In *G. inequilabiatus*, which has 17 sites of 5S rDNA, the sequencing of some clones with classes I and II revealed sequences of the TATA type in NTS, which are associated with the regulation of the expression of the 5S rDNA gene [37].

As a rule, studies investigating this ribosomal gene associate an increase in the number of 5S rDNA sites with the prese-nce of pseudogenes, which would originate through duplication of copies, resulting from the transposition or duplication of genomic DNA, which would be facilitated by the organization tandem of this rDNA family. In addition, locating these sequences in the terminal portion of the

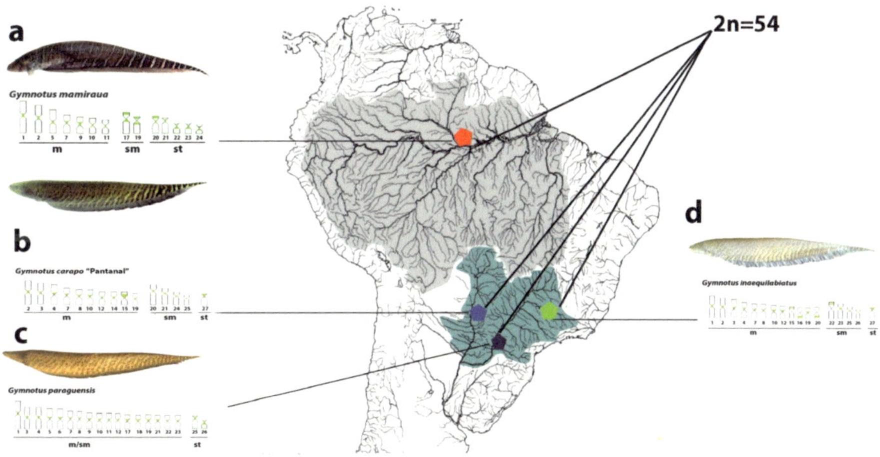

Figure 1.
Location of species with 2n = 54 chromosomes in the Amazon and Paraná-Paraguay basins: a) G. mamiraua *in the Amazon basin; b)* G. carapo*; c)* G. paraguensis *and d)* G. inaequilabiatus*. The pairs with 5S rDNA are represented schematically along with each species.*

chromosomes would facilitate the process of pseudogenization and the association with transposable elements [39].

Although the mechanism of origin and dispersion of these sequences in the 2n = 54 *Gymnotus* genome is still unclear, it is possible to assume that the increase in the number of 5S-like rDNA sites and fixation in the genome accompanies the migratory dispersion of these species. In the case of *Gymnotus* (2n = 54), infection by this transposable element may have probably occurred in a common ancestor for species with 2n = 54 chromosomes, before the final isolation between the Amazon and Paraná-Paraguay basins, since these species are currently found in different river basins. *Gymnotus mamiraua* was restricted to the Amazon basin show the smallest number of sites, 14 pairs, maybe the original satellite sequence like-5S rDNA; however, the isolated species in the Paraná-Paraguay basin experienced divergence from this satellite sequence, differing in number pairs with sequence, such as: *G. inaequilabiatus* (17 pairs), *G. paraguensis* (19 pairs), and *G. carapo* (15 pairs), (**Figure 2**) [13, 22].

Evidence of the connection between South American watersheds is reflected in the evolutionary history of the fish. Other species corroborate the hypothesis of interconnection between these two systems. Migratory species of the genus *Prochilodus* are considered as indicators of this panorama, where the relationship of sister clades of that genus has been useful for understanding the separation of the basins [Magdalena (Orinoco (Amazonas-Paraná)]. The use of molecular clock methodologies, with species in the group, estimates that the isolation between the Amazonas-Paraná basins must have occurred between 2.3–4.1 million years, with an estimated coalescence between strains of 1–3-3 million years [40]. Morphological and molecular analyzes with rheophilic taxa increase the evidence of connection between these basins [41]. The final rupture of the connection between the Paraná-Paraguay and Amazonas-Orinoco basins is inferred as a recent event, with the final uplift of the Andes that changed the course of these basins and isolated them [25, 42], with final separation estimates still uncertain and very peculiar to each group of fish investigated [43].

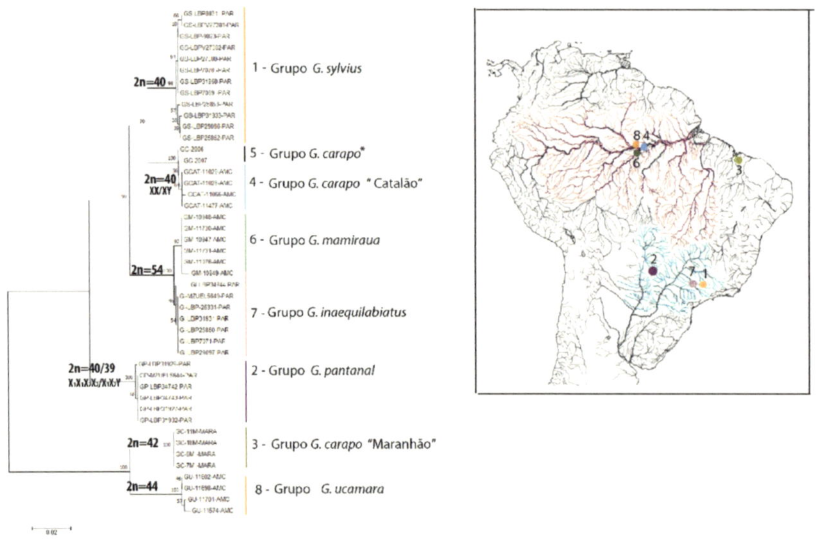

Figure 2.

Topology obtained by the method of clustering neighbors (neighbor joining, NJ) for species of Gymnotus *of the* carapo *clade, using the distance model Kimura-2-parameters (K2P) for the mitochondrial gene COI. The values present in the nodes are the bootstrap values (> 50%) calculated from 1000 replicates and the diploid number superimposed on the tree. The distribution of species is indicated on the map. * absence of coordinates, group not plotted on the map.*

In order to understand the relationships between *Gymnotus*, here, an analysis of a 556 bp (base pairs) COI mitochondrial gene fragment was conducted in 21 specimens from the *carapo* Clade from the central Amazon basin and four more specimens from a population in the Munim River, MA. Sequences deposited in the NCBI database were added to this analysis, with the access numbers described in **Table 1**. The analysis of 51 sequences resulted in eight clades (**Figure 2**). The data revealed a

Species	Voucher	Locality	NCBI accession number
G. sylvius	LBP7069	Alto Paraná	GU.064995.1
G. sylvius	LBP31958	Alto Paraná	GU.701779.1
G. sylvius	LBP7070	Alto Paraná	GU.702209.1
G. sylvius	LBP8831	Alto Paraná	GU.7017821
G. sylvius	LBP27380	Alto Paraná	GU.701778.1
G. sylvius	LBP29096	Alto Paraná	GU.702207.1
G. sylvius	LBP25852	Alto Paraná	GU.701758.1
G. sylvius	LBP31933	Alto Paraná	GU.701767.1
G. sylvius	LBP25853	Alto Paraná	GU.701762.1
G. sylvius	LBPV27382	Alto Paraná	JN.988881.1
G. sylvius	LBPV27381	Alto Paraná	JN.988880.1
G. sylvius	LBP9823	Alto Paraná	GU.701780.1
G. pantanal	MZUEL5644	Alto Paraná	KF.359492.1
G. pantanal	LBP31929	Alto Paraná	GU.701776.1
G. pantanal	LBP34742	Alto Paraná	GU.701775.1
G. pantanal	LBP34743	Alto Paraná	GU.701774.1
G. pantanal	LBP31927	Alto Paraná	GU.701773.1
G. pantanal	LBP31932	Alto Paraná	GU.701763.1
G. carapo Mara	GC7M	Bacia Costeira- Maranhão	XXXXXXXXX
G. carapo Mara	GC9M	Bacia Costeira- Maranhão	XXXXXXXXX
G. carapo Mara	GC11M	Bacia Costeira- Maranhão	XXXXXXXXX
G. carapo Mara	GC18M	Bacia Costeira- Maranhão	XXXXXXXXX
G. carapo Cat	GCAT11029	Amazônia Central	XXXXXXXXX
G. carapo Cat	GCAT 11028	Amazônia Central	XXXXXXXXX
G. carapo Cat	GCAT 11066	Amazônia Central	XXXXXXXXX
G. carapo Cat	GCAT 11477	Amazonia Central	XXXXXXXXX
G. carapo	GC2006	Peru Amazonas	KF533344
G. carapo	GC2007	Peru, Amazonas	KF533345.1
G. mamiraua	GM11078	Amazônia Central	XXXXXXXXX
G. mamiraua	GM11733	Amazônia Central	XXXXXXXXX
G. mamiraua	GM11731	Amazônia Central	XXXXXXXXX
G. mamiraua	GM10947	Amazônia Central	XXXXXXXXX
G. mamiraua	GM10948	Amazônia Central	XXXXXXXXX
G. mamiraua	GM10949	Amazônia Central	XXXXXXXXX
G. mamiraua	GM11730	Amazônia Central	XXXXXXXXX

Species	Voucher	Locality	NCBI accession number
G. inaequilabiatus	MZUEL5649	Alto Paraná	KF.359490.1
G. inaequilabiatus	LBP26331	Alto Paraná	GU.701766.1
G. inaequilabiatus	LBP31931	Alto Paraná	GU.701764.1
G. inaequilabiatus	LBP25850	Alto Paraná	GU.701760.1
G. inaequilabiatus	LBP7071	Alto Paraná	GU.702210.1
G. inaequilabiatus	LBP29097	Alto Paraná	GU.702208.1
G. inaequilabiatus	LBP34744	Alto Paraná	GU.701781.1
G. ucamara	GU11575	Amazônia Central	XXXXXXXXX
G. ucamara	GU11802	Amazônia Central	XXXXXXXXX
G. ucamara	GU11701	Amazônia Central	XXXXXXXXX
G. ucamara	GU11698	Amazônia Central	XXXXXXXXX
G. ucamara	GU11574	Amazônia Central	XXXXXXXXX

Table 1.
Access numbers in the NCBI of the Gymnotus *species of the* carapo *clade used for genetic distance verification through DNA barcode.*

high percentage for the distance between the species identified as *G. carapo* (**Table 1**, **Figure 2**). The sequences of individuals identified as *G. carapo* do not form exclusive groups. Three clades with high rates of genetic distance were observed between *G. carapo* "Catalão" (2n = 40, XX/XY) and *G. carapo* Maranhão (2n = 42), but between *G. carapo* "Catalão" and *G. carapo* do the distance between Amazonas and Peru was 0.02% (**Table 2**). The species *G. mamiraua* forms an exclusive clade with an average intraspecific genetic distance of 0.12% and *G. inaequilabiatus* the average was 0.2% (**Table 3**). These two species showed an interspecific distance of only 0.6%.

The results with the mitochondrial DNA COI obtained here validate the species of *carapo* Clade, with high rates of genetic distance between groups of species identified as *G. carapo*, confirming the probable existence of more than one taxonomic unit in this group. However, this did not occur for two species: *G. mamiraua* from the Amazon basin and *G. inaequilabiatus* from Alto Paraná, for which a low interspecific distance was detected and with a distance of less than 1% was detected, possibly causing recurrent speciation.

Cytogenetic data already pointed to the proximity between these two species, because in addition to sharing the same diploid number, both had many sites of

Intraspecific genetic distance	
Gp 1 - *G. sytlvius*	0.013526931
Gp 2 - *G. Pantanal*	0
Gp 3 - *G. carapo* "Maranhão"	0.000893724
Gp 4 - *G. carapo* "Catalão"	0.000928937
Gp 5 - *G carapo*	0.001858738
Gp 6 - *G.mamiraua*	0.001241471
Gp 7 - *G. inaequilabiatus*	0.002136221
Gp 8 - *G. ucamara*	0.003103677

Table 2.
Intraspecific genetic distance based on mutations of the COI gene, using the K-2-P model.

Interspecific diversity							
	Gp-1	**Gp-2**	**Gp-3**	**Gp-4**	**Gp-5**	**Gp-6**	**Gp-7**
Gp 1 *G. sylvius*							
Gp 2 *G. Pantanal*	0.062						
Gp 3 *G. carapo* "Maranhão"	0.168	0.156					
Gp 4 *G. carapo* "Catalão"	0.030	0.069	0.174				
Gp 5 *G. carapo*	0.028	0.071	0.171	0.002			
Gp 6 *G. mamiraua*	0.048	0.077	**0.181**	0.055	0.053		
Gp 7 *G. inaequilabiatus*	0.046	0.074	0.178	0.053	0.051	**0.006**	
Gp 8 *G. ucamara*	0.179	0.159	0.051	0.183	0.181	0.178	0.174

Table 3.
Matrix of the means of genetic distance (K2P) for the COI gene obtained among the different species of Gymnotus *from* carapo *clade.*

5S rDNA, a condition that is not common in fish. We suggest, by the results of diploid number and dispersion of 5S rDNAr, that the species *G. paraguensis and G. carapo* (2n = 54) from Alto Paraná are also related to the last two already mentioned (**Figure 2**). Thus, what has been verified is that the 5S rDNA has been a potential tool in helping to reconstruct the steps involved in the evolution and biogeographic history of the species of the genus *Gymnotus*, especially the *carapo* Clade, inferred both by the chromosomal mapping and by the molecular analysis of that gene.

2.1 New repetitive sequences studies in the genus *Gymnotus*

The eukaryote genome is characterized by presenting nucleotide sequences with varied arrangements, generally forming two large groups, gene regions and repetitive DNA sequences. In fish of the genus *Gymnotus*, the latter has been associated with heterochromatin. The location of the heterochromatin is reported to be preferentially organized in the centromeric and pericentromeric regions [18, 19, 36, 44].

The prospection of repetitive sequences by the technique of DNA reassociation kinetics (C0t-1) proves to be a safe and fast technique for obtaining copies of highly and moderately repetitive DNA sequences [45]. Thus, it is possible to build libraries and screening repetitive DNAs, and has been used to isolate the highly repeated fraction of the plant genome [46] and animals [47] to significantly expand our knowledge of the organization of their chromosomes.

In the present study, repetitive DNA sequences were isolated by C0t-1 (**Figure 3**) and mapped in two species of electric fish, *G. paraguensis* and *G. sylvius*. Our objective was identifying the heterochromatin compositions and to verify the presence of repetitive sequences originated from transposable elements. Twelve specimens (four females and eight males) of *G. paraguensis* and 21 specimens (seven females and 14 males) of *G. sylvius* collected at Piquiri River, Paraná – Brazil were analyzed.

The isolated probes from *G. paraguensis* had lengths of 473 and 206 bp (**Figure 4a**) and when submitted to BLAST (http://www.ncbi.nlm.nih.gov/blast), clone 2 was found to have 95% identity with microsatellites from *Salmo salar*. Dinucleotide repetitions were observed in this clone. The clones of *G. sylvius* had lengths of 124, 202 and 123 bp (**Figure 4b**). The results of hybridization with total C0t-1 in both species were coincident with heterochromatic sites, according to the description by C banding [18].

The heterochromatin of *Gymnotus* has been reported to be preferentially organized in the centromeric regions [18, 19], as detected in the present study

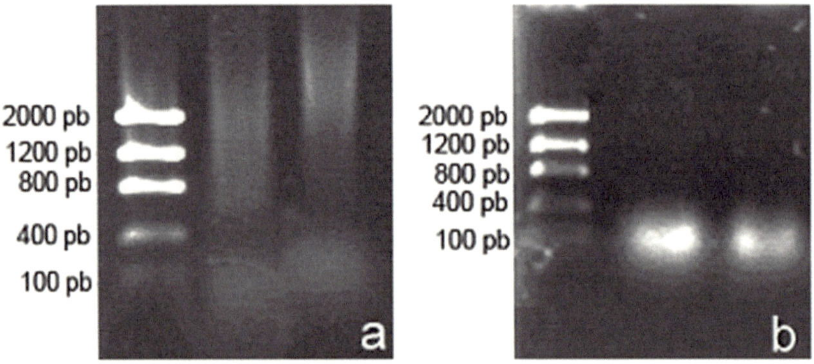

Figure 3.
1% agarose gel electrophoresis of DNA treated for kinetic re-association method ($C_0t - 1$). a) after one minute in autoclave the DNA appears as a trail; b) after treatment with S1 nuclease enzyme the DNA appears with defined length between 100 and 400 bp. (methodology described in complementary material).

a

```
GcotP1 1 GACTGGAAAG CGGGCAGTGA GCGCAACGCA ATTAATGTGA GTTAGCTCAC TCATTAGGCA CCCCAGGCTT
GcotP2 1 ---------- ---------- ---------- ---------- ---------- ---------- ----------
GcotP1   TACACTTTAT GCTTCCGGCT CGTATGTTGT GTGGAATTGT GAGCGGATAA CAATTTCACA CAGGAAACAG
GcotP2   ---------- ---------- ---------- ---------- ---------- ---------- --........
GcotP1   CTATGACCAT GATTACGCCA AGCTCTAATA CGACTCACTA TAGGGAAAGC TTGCATGCCT GCAGGTCGAC
GcotP2   .......... .......... .......... .......... .......... .......... ..........
GcotP1   TCTAGAGGAT CTACTAGTCA TATGGATTGG GGTTGGGGTC TGGGTCCATT GCATAATTGG ATCTAATTCT
GcotP2   .......... ...A...... .......ACA CACACACACA CACACA..C- A..C.CACAC .-.AC.CA.A
GcotP1   TTTTTGTTTG TTTGCAGTCA GTCAA-GTGT CACAAGGGTC AGGGTGCACT TATCAACTCA GTCCAAGAAT
GcotP2   CACACACACA CACA..CA.. C.A.TC.GA. .C.CG..TA. C.A.CT.GAA .CCTGCT.TG ----------
GcotP1   TTGTAGGACA CACCATCAGT AGTGTCCCTC AGAATCCTTG GCTAAAGGCC TGCCAATATT AGTCCAAATG
GcotP2   ---------- ---------- ---------- ---------- ---------- ---------- ----------
GcotP1   CTTTAGGGGT AGATCGGATC CCCGGGTACC GAGCTCGAAT CCTGCCTTTT GGGG 474
GcotP2   ---------- ---------- ---------- ---------- ---------- ---- 474
```

b

```
G.s1   ---------- ---------- ---------- ---------- ---------- ---------- ----------
G.s2   TCATATGGAT TTCCTGCTGA AGCAGGACAC GGCTGTTTCC TGCTGGCGTC CTGGCTTCCT GCACACTGGA
G.s3   ---------- ---------- ---------- ---------- ---------- ---------- --------GA

G.s1   ---------- TTAACTTCCT GACCGATCTA GAGGA-CTAC TAGTCATATG GATCTGGCTG TGCATCACTG
G.s2   CTGGTTCCCT C..T..AATC AGT.AC.G.. C.CTG--G.. AG..TCCC.C T....AATCA GT..CTG.AA
G.s3   AAGGGCACCT C.G....GAG CGT....T.T TGT..T.CT. C.AAAGGGC. ..G.CTATG. AAA.A...CA

G.s1   GCTGTGC-AT CACTCCTTGT GTATCACTGG CTGTGTATCA CTGGCTGTAC ACCCACTCCT TTGATC 206
G.s2   A...GA.TGG TT.C.TC.A. C....GGATC .CCG.GTA.C GA.CTC.A.T C..TG..TT. GG.G-- 206
G.s3   ..AAG..G.. .GG...CC.G .ACC.GAA.C T..GAATG.G GG..GG..TG G--------- ------ 206
```

Figure 4.
a) Alignment of two isolated clones of G. paraguensis, *b) alignment of three clones of* G. sylvius.

(**Figure 5a** and **b**), and inmost species of the Gymnotidae family [36, 44]. Furthermore, heteromorphisms in length between homologous chromosomes of the NOR regions were observed in *G. paraguensis*, indicating differential accumulations of heterochromatin regions (**Figure 5b**). However, when the probes of one species were hybridized with the other (cross-FISH) no positive hybridization signals were observed (data not shown).

According to Charlesworth et al. [48] and Topp and Dave [49], the regions located nearest to the centromere show fast evolutionary rates due to low recombination, initiating the accumulation of repetitive DNA sequences, which explains its specificity. This association between heterochromatin and repetitive sequences is fundamental to the organization of important chromosomal structures such as the centromere. In a study with *Oreochromis niloticus* using the GISH (Genomic *in situ* Hybridization) methodology, the heterochromatin present in pericentromeric regions was found to be species-specific [50] and composed of repetitive and transposable elements [51–53].

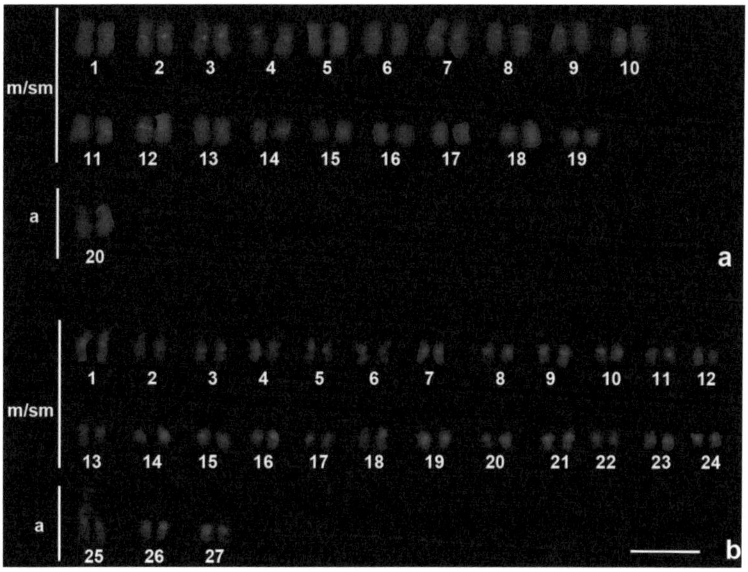

Figure 5.
Fluorescent in situ *hybridization (FISH) performed with probes isolated for kinetic re-association of DNA (C_0t–1). a)* G. sylvius *karyotype; b)* G. paraguensis *karyotype. Observe in b the heteromorphism of nucleolus organizer regions in pair 25. Bar = 10 μm.*

The repetitive elements isolated from *G. sylvius* and *G. paraguensis* in the present study were located in the pericentromeric region, coincident with the heterochromatin observed by C banding. High levels of specificity of the isolated probes and of the species' genomes were assumed because no signals were observed in crossed FISH analyses - confirming the heterogeneous composition of the heterochromatin of these species.

Sequencing analyses showed exclusive sequences for both species, and although repetitive elements in the heterochromatin regions are present in distant eukaryotes groups such as *Drosophila* and plants [54, 55] as important structural regions of the genome, the structural functions of these sequences in *G. sylvius* and *G. paraguensis* are not yet known.

The Y chromosome of *Eigenmannia virescens* isolated by Henning et al. [56] had large amounts of heterochromatin and physical mapping with Y chromosome probe was performed with closely related species without differentiated sex chromosome systems, and the probes hybridized only in the centromeric and telomeric regions.

In addition to the two species analyzed in the present study, two other species of Gymnotiformes have had their repetitive DNA sequences analyzed and described. Claro [57] isolated the repetitive sequences of *G. sylvius* and *G. carapo* by enzymatic digestion with *Alu*I and *Hae*III. The isolated fragments with 300 bp showed dispersed distributions in both the species and similar locations using both enzymes. Furthermore, a transposable SINE element that labeled different regions in *G. carapo* was identified; all the sequences showed disperse labeling that was not coincident with heterochromatin regions, suggesting an important function in the evolution and organization of non-codifying DNA regions [58].

More recently, the publication of Satellitome results has been awaited, a global study by NGS and bioinformatics of all satellite sequences of *Gymnotus* species from the main clades. According to previous data released by the authors, the massive characterization of satellite DNA families processes of genetic differentiation and the dynamics of these sequences among the representatives stand out [59], which corroborates our findings in the present work.

3. Conclusions

Molecular studies with the multigene family 5S rDNA in electric fish (Gymnotidae) have advanced a lot in recent years. The chromosomal location and distribution have been particularly interesting, since all species of the genus *Gymnotus*, analyzed so far, show great variability. Its distribution on the chromosomes of species with 2n = 54 shows to be a biogeographic marker, suggesting that the speciation of this group was recent due to migratory expansion. On the other hand, the species analyzed in the present study showed repetitive sequences composed of microsatellite replications species-specific, present around the centromeres. Although the location of the heterochromatin in *Gymnotus* was conserved, they had different constitutions and could represent important evolutionary markers for cytotaxonomic studies of this group.

Acknowledgements

The current study was supported in part by INCT ADAPTA II funded by CNPq - Brazilian National Research Council (465540/2014-7). The authors are grateful to Miguel Airton Carvalho for his field assistance. The author M.D.S. received a scholarship from the *Conselho Nacional de Desenvolvimento Científico e Tecnológico* (CNPq) (grant 160155/2018-5).

Conflict of interest

The authors declare no conflict of interest.

A. Appendices

A.1 Methodology for obtaining cytochrome oxidase sequences

For the Cytochrome oxidase I gene, five specimens of *G. carapo* "Catalão", four specimens of *G. carapo* "Maranhão", four specimens of *G. ucamara* and seven specimens of *G. mamiraua* were sequenced, using the primers BOL-COIfishF1 (5'TCAACYAATCAYAAAGATATYGGCAC′3′) and BOL-COI-fishR1 (5′-ACTTCYGGGTGRCCRAARAATCA- 3′) [60]. Genomic DNA extraction was performed using the chloroform-phenol method [61]. The PCR reactions were performed in a final volume of 25 μL, containing genomic DNA (100 ng), 10x buffer with 1.5 mM $MgCl_2$, Taq DNA polymerase (5 U/μL), dNTPs (1 mM), pair of primers (5 pM) and Milli-Q water. The conditions for amplification were: 1 min 95° C, followed by 30 cycles of 1 min at 94° C, 1 min at 59° C, 1 min 30 sec at 72° C and final extension of 5 min at 72° C.

Sequences of *Gymnotus* species for the Clado *carapo*, available in the NCBI database, from the Paraná-Paraguay basin were added to our analysis: *G. sylvius*, *G. inaequilabiatus*, *G. pantanal* and *G. carapo* "Pantanal" (**Table 1**).

The calculation of intra and interspecific distance was performed using the Mega 5 software [62] with the Kimura-2-parameters evolutionary model. The identification was carried out according to the protocol established by DNA barcoding through the Neighbor Joining (NJ) analysis [63], which consists of looking for the tree with the lowest total sum of branches, using Kimura-2-parameters as an evolutionary model (K2P) [64]. The topology confidence test was performed with bootstrap analysis, containing 1000 replicates. Such analyzes were performed using the Mega 5 software [62].

The sequences were aligned in the program Clustal W [65], using the editor BioEdit 7.0 [66], were submitted to BLAST in the NCBI database (http://www.ncbi.nlm.nih.gov).

A.2 Methodology for obtaining cytogenetic material

Mitotic chromosomes were obtained according to the protocol described by Bertollo et al. [67]. Genomic DNA extraction was performed using the chloroform-phenol method [61]. Repetitive DNA probes were obtained using the C_0t-1 DNA technique as described by Zwick et al. [45] and adapted by Vicari et al. [68]. This technique is based on re-association kinetics and enzymatic digestion by S1 nuclease. The probes were labeled by nick translation with digoxigenin 11 dUTP (Roche®) and the signals were recognized by anti-digoxigenin-rhodamin (Roche®). The hybridization techniques followed the Pinkel et al. [69] protocol, with 77% of stringency (2.5 ng/μL of each probe, 50% of deionized formamide, 10% dextran sulphate, 2X SSC at 37°C for 18 hours). The chromosomes were counterstained with DAPI in Vectashield medium (Vector®). The chromosomal preparations were analyzed by epifluorescence microscopy using an Olympus BX41® microscope fitted with a CCD Olympus DP-71® digital camera. Image capture was performed using DP controler® software (Olympus). The probes with positive hybridization signals were purified and cloned using the pMOSBlue blunt ended RPN5110 cloning kit (Amersham Biociences®) for subsequent sequencing. The samples were sequenced in an ABI-PRISM 3100 Genetic Analyzer automatic sequencer at the ACTGene laboratory (Centro de Biotecnologia, UFRGS, Porto Alegre, RS).

The sequences were aligned and edited using the CLUSTAL W program [65] using the following parameters: weights 6.66 and 10.0 for opening and extension of gaps, respectively, for pairwise alignments, and weights 10.0 and 15.0 for opening and extension of gaps in the multiple alignments respectively. Clustering analysis was performed using the parsimony method on the PAUP program v. 4.0b10 [70]. The isolated fragments of repetitive DNA sequences (C_0t-1) from *G. sylvius* and *G. paraguensis* had lengths between 100 and 400 bp, respectively (**Figure 5**). These fragments were used as probes for fluorescent *in situ* hybridization (FISH) and were found to label most of the pericentromeric regions of the chromosomes of *G. sylvius* (**Figure 4a**) and all of the chromosomes of *G. paraguensis* (**Figure 4b**).

Author details

Maelin da Silva[1*], Daniele Aparecida Matoso[2], Vladimir Pavan Margarido[3], Eliana Feldberg[4] and Roberto Ferreira Artoni[5]

1 Department of General Biology, State University of Ponta Grossa, Paraná, Brazil

2 Department of Genetics, Institute of Biological Sciences, Federal University of Amazonas, Manaus, Brazil

3 Center for Biological and Health Sciences, Genetics Laboratory, State University of Western Paraná, Paraná, Cascavel, Brazil

4 National Amazon Research Institute, Manaus, Brazil

5 Department of Structural Biology, Molecular and Genetics, State University of Ponta Grossa, Ponta Grossa, Brazil

*Address all correspondence to: maelinbio@gmail.com

References

[1] Heslop-Harrison JSP, Schwarzacher T. Organisation of the plant genome in chromosomes. Plant J. 2011;66:18-33. DOI:10.1111/j.1365-313X.2011.04544.x

[2] Jurka J, Kapitonov VV, Pavlicek A, Klonowski P, Kohany O, Walichiewicz J. Repbase Update, a database of eukaryotic repetitive elements. Cytogenet. Gen. Res. 2005; 110:462-467. DOI: 10.1159/000084979.

[3] Grady DL, Ratliff RL, Robinson DL, McCanlies EC, Meyne J, Moyzis RK. Highly conserved repetitive DNA sequences are present at human centromeres. Proc Natl. Acad. Sci. USA. 1992; 89:1695-1699.

[4] Schueler MG, Higgins AW, Rudd MK, Gustashaw K, Willard H. Genomic and genetic definition of a functional human centromere. Science. 2001;294:109-115. DOI: 10.1126/science.1065042.

[5] Garrido-Ramos MA, Jamilena M, Lozano R, Ruiz Rejón C, Ruiz Rejón M. Cloning and characterization of a fish centromeric satellite DNA. Cytogenetic and Genome Research. 1994;65:233-237. DOI:10.1159/000133637.

[6] Martins C, Ferreira IA, Oliveira C, Foresti F, Galetti Jr. PM. A tandemly repetitive centromeric DNA sequence of the fish *Hoplias malabaricus* (Characiformes: Erythrinidae) is derived from 5S rDNA. Genetica. 2006;127: 133-141. DOI: 10.1007/s10709-005-2674-y.

[7] Mazzuchelli J, Martins C, Genomic organization of repetitive DNAs in the cichlid fish *Astronotus ocellatus*. Genetica. 2009;136: 461-469. DOI: 10.1007/s10709-008-9346-7.

[8] CioffiMB, Martins C, Bertollo LAC. Evidence spreading of associated transposable elements and ribosomal DNA in the fish *Erythrinus erythrinus*. Implications for genome change and karyoevolution in fish. BMC Evolutionary Biology. 2010;10: 271-280. DOI: 10.1186/1471-2148-10-271.

[9] Silva M, Ribeiro ED, Matoso DA, Sousa LM, Hrbek T, Rapp Py-Daniel L. Feldberg, E. Chromosomal polymorphism in two species of *Hypancistrus* (Siluriformes: Loricariidae): an integrative approach forunderstanding their biodiversity. Genetica. 2014;142(2): 127-139. DOI: 10.1007/s10709-014-9760-y.

[10] Cioffi MB, Bertollo LAC. Initial steps in XY chromosome differentiation in *Hoplias malabaricus* and the origin of an X_1X_2Y sex chromosome system in this fish group. Heredity. 2010;105:554-561. DOI:10.1038/hdy.2010.18.

[11] Borba RS, Silva EL, Parise-Maltempi PP. Chromosome mapping of retrotransposable elements *Rex1* and *Rex3* in *Leporinus* Spix, 1829 species (Characiformes: Anostomidae) and its relationships among heterochromatic segments and W sex chromosome. Mobile Genet Elements. 2013; 3(6):e27460. DOI: 10.4161/mge.27460.

[12] Mariotto S, Centofante L, Artoni RF, Vicari MR, Moreira-Filho O. Chromosomal diversification in ribosomal DNA sites in *Ancistrus* species (Loricariidae: Ancistrini) from three hydrographic basins of Mato Grosso, Brazil. Comparative Cytogenetics. 2011;5:289-300. DOI:10.1007/s1116001192159.

[13] Silva M, Matoso DA, Artoni RF, Feldberg E. Karyotypic diversity and evolutionary trends in Neotropical electric fish of the genus *Gymnotus* (Gymnotiformes: Gymnotidae). Zebrafish. 2019;16:308-320. DOI: 10.1089/zeb.2018.1716.

[14] Vicari MR, Artoni RF, Moreira-Filho O, Bertollo LAC. Colocalization of repetitive DNAs and

silencing of major rRNA genes. A case report of the fish *Astyanax janeiroensis*. Cytogenetics Genome Research. 2008;122:67-72. DOI: 10.1159/ 000151318.

[15] Ferreira IA, Martins C. Physical chromosome mapping of repetitive DNA sequences in Nile tilapia *Oreochromis niloticus*: evidences for a differential distribution of repetitive elements in the sex chromosomes. Micron. 2008;39:411-418. DOI:10.1016/ j.micron. 2007.02.010

[16] Silva FA, Guimarães EC, Carvalho NDM, Ferreira AMV, Schneider CH, Carvalho-Zilse, GA, Feldberg E, Gross MC. Transposable DNA elements in amazonian fish: from genome enlargement to genetic adaptation to stressful Environments. Cytogenetics Genome Research. 2020;10. DOI: 10.1159/000507104

[17] Almeida-Toledo LF, Foresti F, Pe'Quignot EV, Daniel-Silva MFZ. XX/ XY sex chromosome system with X heterochromatinization: an early stage of sex chromosome differentiation in the Neotropic electric eel *Eigenmannia virescens*. Cytogenet. Cell Genet. 2001;95:73-78. DOI: 10.1159/000057020.

[18] Margarido VP, Bellafronte E, Moreira-Filho O. Cytogenetic analysis of three sympatric *Gymnotus* (Gymnotiformes, Gymnotidae) species verifies invasive species in the Upper Paraná River basin, Brazil. Journal of Fish Biology. 2007;70:155-164. DOI: 10.1111/ j.1095-8649.2007.01365.x.

[19] Milhomem SSR, Pieczarka JC, Crampton WGR, Silva DS, Souza ACP, Carvalho JR, Nagamachi CY. Chromosomal evidence for a putative cryptic species in the *Gymnotus carapo* species-complex (Gymnotiformes, Gymnotidae). BMC Genetics. 2008; 9:75. DOI: 10.1186/1471-2156-9-75.

[20] Nagamachi CY, Pieczarka JC, Milhomem SSR, O'Brien PCM, Souza ACP, Ferguson-Smith MA. Multiple rearrangements in cryptic species of electric knifefish, *Gymnotus carapo* (Gymnotidae, Gymnotiformes) revealed by chromosome painting. BMC Genetics. 2010;11:2-9. DOI: 10.1186/ 1471-2156-11-28.

[21] Utsunomia R, Melo S, Scacchetti PC, Oliveira C, Machado MA, Pieczarka JC, Nagamachi CY, Foresti F. Particular chromosomal distribution of micro-satellites in five species of the genus *Gymnotus* (Teleostei, Gymnotiformes). Zebrafish. 2018;15:398-403. DOI:10.1089/ zeb.2018.1570.

[22] Silva M, Barbosa P, Artoni R F, Feldberg E. Evolutionarydynamics of 5S rDNA and recurrent association of transposable elementsinelectric fish of the family Gymnotidae (Gymnotiformes): the case of *Gymnotus mamiraua*. Cytogenet. Genome Res. 2016;149:297-303. DOI: 10.1159/ 000449431.

[23] Fernades-Matioli FMC, Almeida-Toledo LF. A molecular phylogenetic analysis in *Gymnotus* species (Pisces: Gymnotiformes) with inferences on chromosome evolution. Caryologia 2001;54(1):23-30.

[24] Silva EB, Margarido VP. An $X_1X_1X_2X_2/X_1X_2Y$ multiple sex chromosome system in a new species of the genus *Gymnotus* (Pisces, Gymnotiformes). Environmental Biology of Fishes. 2005;73:293-297. DOI: 10.1007/ s10641-005-2144-5.

[25] Lundberg JG, Marsall LG, Horton B, Malabarba CSL, Wesselingh F. The stage for Neotropical fish diversification: A history of tropical South American rivers. In: Malabarba LR, Reis RE, Vari RP, Lucena CAS, Lucena ZMS. (eds) Phylogeny and classification of

neotropical fishes. EDIPURCS, Porto Alegre, RS, 1998. p. 13-48.

[26] Nelson, J. Fishes of the World. Hoboken, NJ, John Wiley and Sons. 2006.

[27] Crampton WGR, Thorsen DH, Albert JS. Three new species from a diverse and sympatric assemblage of the electric fish *Gymnotus* (Ostariophysi: Gymnotidae) in the lowland Amazon Basin, with notes on ecology. Copeia. 2005:82-99.

[28] Campos-da-Paz R. Family Gymnotidae. In: Reis RE, Kullander SO, Ferraris Jr. CJ. (eds.). Check List of the Freshwater Fishes of South and Central. EDIPURCS, Porto Alegre, 2003. p. 483-486.

[29] Albert JS, Miller RR. *Gymnotus maculosus*: a new species of electric fish from Middle America (Teleostei: Gymnotoidei), with a key to the species of *Gymnotus*. Proceedings Biological Society Washington. 1995;108:662-678.

[30] Albert JS, Campos-da-Paz R. Phylogenetic systematics of America knifefishs: a review of the available data. In: Malabarba L, Reis RE, Vari RP, Lucena CAS, Lucena ZMS. (ed) Phylogeny and Classification of Neotropical Fishes. EDIPUCRS, Porto Alegre, 1998. p. 419-446.

[31] Oyakawa OT, Akama A, Mautari KC, Nolasco JC. Peixes de riachos da Mata Atlântica nas Unidades de Conservação do Vale do Rio Ribeira de Iguape no Estado de São Paulo. São Paulo. Ed. Neotrópica. 2006.

[32] Alves-Gomes JA. The Mitochondrial Phylogeny of the South American Electric Fish (Teleostei: Gymnotiformes) and an Alternative Hypothesis for the Otophysan Historical Biogeography. In: Poyato-Ariza FJ, Grande T, Diogo R (ed.) Teleostean Fish Biology: Gonorynchiformes and Ostariophysan Relationships: A Comprehensive Review. Enfield, Science Publishers, 2009. p. 517-565.

[33] Milhomem SSR, Crampton WGR, Pieczarka JC, Shetka GH, Silva DS Nagamachi, CY. *Gymnotus capanema*, a new species of electric knife fish (Gymnotiformes, Gymnotidae) from Eastern Amazonia, with comments on an unusual karyotype. Journal Fish Biology. 2012;80:802-815. DOI:10.1111/j.1095-8649.2012.03219.x.

[34] Milhomem SSR, Pieczarka JC, Crampton WR, Souza ACP, Carvalho Jr. JR, Nagamachi CY. Differences in karyotype between two sympatric species of *Gymnotus* (Gymnotiformes:Gymnotidae) from the eastern Amazon of Brazil. Zootaxa. 2007;1397: 55-62. DOI:10.1186/1471-2156-9-75.

[35] Silva M, Matoso DA, Vicari MR, Almeida MC, Margarido VP, Artoni RF. Repetitive DNA and meiotic behavior of sex chromosomes in *Gymnotus pantanal* (Gymnotiformes, Gmynotidae). Cytogenetic and Genome Research. 2011;135:143-149. DOI:10.1159/000330777.

[36] Silva M, Matoso DA, Vicari MR, Almeida MC, Margarido VP, Artoni RF. 2011. Physical mapping of 5S DNAr in two species of Knifefishes: *Gymnotus pantanal* and *Gymnotus paraguensis* (Gymnotiformes). Cytogenetic Genome Research. 2011;134:303-307. DOI:10.1159/000328998.

[37] Scacchetti PC, Alves JC, Utsunomia R, Claro FL, Almeida-Toledo LF, Oliveira C, Foresti F. Molecular characterization and physical mapping of two classes of 5S rDNA in genomes of *Gymnotus sylvius* and *G. inaequilabiatus* (Gymnotiformes, Gymnotidae). Cytogenetics Genome Research. 2011;136:1-7. DOI:10.1159/000335658.

[38] Scacchetti PC, Pansonato-Alves JC, Utsunomia R, Oliveira C, Foresti F. Karyotypic diversity in four species of the genus *Gymnotus* Linnaeus, 1758 (Teleostei, Gymnotiformes, Gymnotidae): physical mapping of ribosomal genes and telomeric sequences. Comparative Cytogenetics. 2012;5:223-235. DOI:10.3897/CompCytogen.v5i3.1375.

[39] Sochorová J, Garcia S, Gálvez F. et al. Evolutionary trends in animal ribosomal DNA loci: introduction to a new online database. Chromosoma. 2018;127:141-150. DOI:10.1007/s00412-017-0651-8.

[40] Sivasundar A, Eldregde B, Orti G. Population structure and biogeography of migratory freshwater fishes (*Prochilodus*: Characiformes) in major South American rivers. Molecular Ecology. 2001;10:407-417. DOI:10.1046/j.1365-294x.2001.01194.x.

[41] Ribeiro AC, Jacob RM, Silva RRRS, Lima FCT, Ferreira DC, Ferreira KM, Mariguela TC, Pereira LHG, Oliveira C. Distribuitions and phylogeographic data of rheophilic freshwater fishes provide evidence on the geographic extension of a Central – Brazilian Amazonian palaeo plateau in the area of the present day Pantanal wetland. Neotropical Ichthyology.2013;11:319-326. DOI: 10.1590/S1679-62252013000200010.

[42] Hoorn, C.; Wessilingh, F.P. Introduction: Amazonia, landscape and species evolution. In: Hoorn, C.; Wessilingh, F.P. (Eds). Amazonia: landscape and species evolution. A look into the past. Wiley-Blackwell, Oxford, 2010. p.1-9.

[43] Montoya-Burgos JI. Historical biogeography of the catfish genus *Hypostomus* (Siluriformes: Loricariidae), with implications on the diversification of Neotropical ichthyofauna. Molecular Ecology. 2003;12:1855-1867. DOI: 10.1046/j.1365-294X.2003.01857.x.

[44] Silva M, Matoso DA, Vicari MR, Almeida MC, Margarido VP, Artoni RF. Physical mapping of 5S rDNA in two species of knifefishes: *Gymnotus pantanal* and *Gymnotus paraguensis* (Gymnotiformes). Cytogenetic and Genome Research. 2011;134,303-307. DOI:10.1159/000328998.

[45] Zwick MS, Hanson RE, McKnight TD, Islam-Faridi HM, Stelly DM, Wing RA, Price JH. A rapid procedure for the isolation of C_0t-*1* DNA from plants. Genome. 1997;40:138-142.

[46] Hřibová E, Doleželová M, Town CD, Macas J, Doležel J. Isolation and characterization of the highly repeated fraction of the banana genome. Cytogenet. Genome Res. 2007;119:268-274. DOI:10.1159/000112073.

[47] Lorscheider CA, Oliveira JIN, Thais AD, Nogaroto V, Martins-Santos IC, Vicari MR. Comparative cytogenetics among three sympatric *Hypostomus* species (Siluriformes: Loricariidae): An evolutionary analysis in a high endemic region. Braz. Arch. Biol. Technol. 2018;61:14. DOI: 10.1590/1678-4324-2018180417

[48] Chalesworth B, Sniegowski P, Stephan W. The evolutionary dynamics of repetitive DNA in eukaryotes. Nature. 1994;371:215-220.

[49] Topp CN, Dawe RK. Reinterpreting pericentromeric heterochromatin. Curr. Opin. Plant. Biol. 2006;9:647-653. DOI:10.1016/j.pbi.2006.09.008.

[50] Valente GT, Schneider CH, Gross MC, Feldeberg E, Martins C. Comparative cytogenetics of cichlid fishes through genomic in-situ hybridization (GISH) with emphasis on *Oreochromis niloticus*. Chromosome Resarch. 2009. DOI:10.1007/s10577-009-9067-5.

[51] Oliveira C, Wright JM. Molecular cytogenetic analysis of heterochromatin in the chromosomes of tilapia, *Oreochromis niloticus* (Teleostei:

Cichlidae). Chromosome Research. 1998;6:205-211.

[52] Oliveira C, Joyce S, Chew K, Porto-Foresti F, Dobson MJ, Wright JM. 1999. A LINE2 repetitive DNA sequence from the cichlid fish, *Oreochromis niloticus:* sequence analysis and chromosomal distribution. Chromosoma 1999;108:457-468.

[53] Grewal SIS, Jia S. Heterochromatin revisited. Nat. Rev. Genet. 2007;8:35-46. DOI:10.1038/nrg2008.

[54] Pimpinelli S, Berloco M, Fanti L, Dimitri P, Bonaccors S, Marchettib E, Caizzi R. Transposable elements are stable structural components of *Drosophila melanogaster* heterochromatin. Proc. Natl. Acad. Sci. USA. 1995;92:3804-3808.

[55] Presting GG, Malysheva L, Fuchs J, Schubert I. A Ty3/gypsy retrotransposon-like sequence localizes to the centromeric regions of cereal chromosomes. Plant. J. 1998;6:721-728. DOI:10.1046/ j.1365-313x.1998.00341.x

[56] Henning F, Trifonov V, Almeida-Toledo LF. Use of chromosome microdissection in fish molecular cytogenetic. Genetics and Molecular Biology. 2008:31:279-281. DOI:10.1590/ S1415-47572008000200022.

[57] Claro FL. *Gymnotus carapo* e *Gymnotus sylvius* (Teleostei: Gymnotidae): Uma abordagem Citogenética-Molecular. [Dissertation]. São Paulo - Brasil: Universidade de São Paulo; 2008.

[58] Buldyrev SV, Goldberger AC, Harlin S, Mategna RN, Matsa ME, Peng CK, Simmons M, Stanley HE. Long-range correlation proprieties of coding and noncoding DNA sequences. Gene Bank. Analysis. Phys. Rev. E. 1995; 51:5084-5091.

[59] Melo S. Análises cromossômicas e genômicas aplicadas ao estudo evolutivo e estrutural dos satelitomas em espécies do gênero *Gymnotus* (Teleostei: Gymnotiformes). [Thesis]. Botucatu –SP: Universidade Estadual Paulista "Júlio de Mesquita Filho", Instituto de Biociências de Botucatu; 2020.

[60] Ward RD, Zemlak TS, Innes BH, Last PR, Hebert PDN. DNA barcoding Australia's fish species. Philosophical Transactions of the Royal Society Biological Sciences. 2005;360:1847-1857. DOI:

[61] Sambrook J, Russell DW. Molecular cloning: a laboratory manual. Cold Spring Harbor Laboratory Press, New York. 2001. p. 2-5.

[62] Tamura K, Nei M, Kumar S. MEGA4: Molecular Evolutionary Genetics Analysis (MEGA) Software Version 4.0. Molecular Biology and Evolution. 2007; 24, 1596-1599. DOI:10.1093/molbev/msm092.

[63] Saitou N, Nei M. The Neighbor-joining method: a new method for reconstructing phylogenetic trees. Molecular Evolution. 1987;4:406-425.

[64] Kimura M. A simple method for estimating evolutionary rates of base substitutions through comparative studies of nucleotide sequences. Journal of Molecular Evolution. 1980;16:111-120.

[65] Thompson JD, Higgins DG, Gibson TJ. CLUSTAL W: improving the sensitivity of progressive multiple sequence alignment through sequence weighting, position-specific gap penalties and weight matrix choice. Nucleic Acids Research. 1994;22:4673-4680.

[66] Hall TA. BioEdit: a user-friendly biological sequence alignment editor and analysis program for Windows 95/96/NT. Nucleic Acids Symposium Series. 1999;41:95-98.

[67] Bertollo LAC, Takahash CS, Moreira-Filho O. Cytotaxonomic

considerations on *Hoplias lacerdae* (Pisces, Erythrinidae). Rev. Bras. Genet. 1978;1(2):103-120.

[68] Vicari MR, Nogaroto V, Noleto RB, Cestari MM, Cioffi MB, Almeida MC, Moreira-Filho O, Bertollo LAC, Artoni RF. Satellite DNA and chromosomes in Neotropical fishes: methods, applications and perspectives. Journal of Fish Biology. 2010;76:1094-1116. DOI: 10.1111/j.1095-8649.2010.02564.x.

[69] Pinkel D, Straume T, Gray JW. Cytogenetic analysis using quantitative, high-sensitivity, fluorescence hybridization. Proc. Natl. Acad. Sci. USA. 1986;83:2934-2938.

[70] Swofford DL. PAUP Phylogenetic analysis using parsimony (and other methods). Version 4. Sinauer Associates, Sunderland, Massachusetts. 2002. DOI:10.1111/j.0014-3820.2002.tb00191.x

Chapter 3

Heterochromatin Dynamics in Response to Environmental Stress in Amazonian Fish

Daniele Aparecida Matoso, Maelin da Silva, Hallana Cristina Menezes da Silva, Eliana Feldberg and Roberto Ferreira Artoni

Abstract

Transcriptionally inactive portions of genomic DNA, condensed with histones and architectural proteins, are known as heterochromatic regions, often positive C band. The advent of epigenetics and new methodological approaches, showed that these regions are extremely dynamic and responsive to different types of environmental stress. The relationship of the constitutive heterochromatin with the transposable elements inactivation, especially from the Rex family, seems to be a frequent condition in fish. In this manuscript we review the existing knowledge of the nature and function of these genomic regions, based on species-based studies, with a focus on species of fish from the Amazon region.

Keywords: environmental stressors, transposable elements, adaptive response, *Colossoma*, *Hypancistrus*

1. Introduction

The genomic DNA of eukaryotic organisms combines with histone proteins to form chromatin. Chromatin is classified into two forms: euchromatin (de-condensed region, rich in genes and transcriptionally active) and heterochromatin (condensed, transcriptionally silent) [1–3]. This early classification was based on differing dye-responses and condensation profiles [4]. Heterochromatin, in turn, can be classified into constitutive heterochromatin and facultative heterochromatin, the former being preferably assembled in regions that house repetitive elements, such as satellite DNA and transposable elements [2, 5, 6]. The latter is preferentially assembled in genes related to the regulation of organismal development. The idea that the material is strongly related to the heterochromatinization of one of the X sex chromosomes in female mammals is known as the Lyon hypothesis [7].

Recent studies have shown that both constitutive and facultative heterochromatin are regulated dynamically and are responsive to various stressful stimuli. It is also known that while these changes in chromatin structure can potentially help organisms adapt to new environments, they can also produce aberrant phenotypes [8, 9] and diseases in humans [10, 11]. They are also related to the aging process [12]. In this review, we will discuss the dynamics of heterochromatin localization, obtained from different studies in fish from the Amazon region (**Figure 1**).

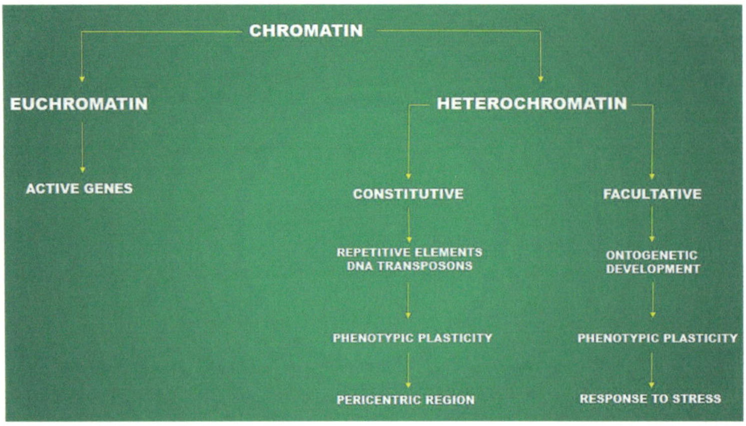

Figure 1.
General distribution of chromatin in eukaryotos. This schematization compiles information from several studies. Here, euchromatin is represented as the portion of chromatin with transcriptionally active genes. Meanwhile, heterochromatin can be divided into two groups depending on constitution and activity. The constitutive herochromatin is composed of repetitive sequences and transposable elements, distributed in pericentromeric and terminal region of chromosomes that occasionally participate in phenotypic plasticity, through the regulation of chromatin and dispersion of transposable elements, uneven crossing-over or rearrangements. On the other hand, the portion of facultative heterochromatin, regulated during ontogenetic development, is related to the response of environmental stress or laboratory simulated, allowing plasticity and environmental adaptation.

2. Molecular characteristics of heterochromatin

Facultative heterochromatin is traditionally considered to have a more plastic structure than constitutive heterochromatin.

At the molecular level, facultative heterochromatin is composed of transcriptionally silent chromatin regions, which condense or decondense, thus allowing transcription in temporal and spatial contexts [13]. Therefore, facultative heterochromatin formation appears to be directly linked to the different isoforms of histone H1. Here, under the facultative heterochromatin formation model proposed by [14], the different H1 isoforms can take on unique functions via specific changes in chromatin structure.

The concept of facultative heterochromatin was developed to explain the phenomenon of dose compensation in mammalian cells. The X chromosome, inactive in female mammalian cells, is subject to a monoallelic suppression of genes that depends on numerous chromatin modifiers, resulting in extensive condensation [15]. This process involves non-coding RNA (ncRNA) called X_{ist}, which is exclusively expressed by the inactive X chromosome. This ncRNA is responsible for post-translational modifications of histones, among which the most common forms include H4 lysine 20 methylation (H4K20me), H3 lysine 27 trimethylation (H3K27me3), and H3 lysine 9 methylation [16, 17]. Another important variation is the incorporation of the histone macro H2A, while Polycomb (PcG), Polycomb repressor complex 2 (PRC2), and PRC1 are also involved in the process. It is generally considered that Polycomb (PcG) proteins play a central role in the formation of facultative heterochromatin, with the most powerful histone modification being the methylation of H3 lysine 27 (H3K27me) [13].

Hypoacetylation of histone tails, binding of the HP1 protein with H3K9me2/3, and ubiquitination of H2A lysine 119, as well as the presence of histone macro H2A, appear to provide a molecular signature for the composition of facultative heterochromatin [13]. The fact that facultative heterochromatin is maintained across cell generations, with PcG proteins, ncRNA, and trans- acting transcription factors as participants, shows that this form of heterochromatin may be largely responsible for phenotypic differences, which can be inherited or arise spontaneously in response to environmental challenges or ontogenetic development [13, 15].

Some studies have shown that the HP1 protein undergoes changes during the cell differentiation process [18]. Such changes are considered indicative of a highly conserved regulatory mechanism for the assembly of heterochromatin in response to environmental stress [12]. The phosphorylation function of HP1 is flexible, allowing responses to various stimuli and permitting more finessed cellular adaptation to environmental changes [12]. This suggests that epigenetic changes, mediated by heterochromatin, constitute a quick and efficient mechanism for generating flexible cell tolerance responses to environmental stress [12].

It has been established that the composition and formation of the constitutive heterochromatin are similar to that of the facultative heterochromatin regions, with these regions being hypoacetylated and containing histone H3 with hypermethylated lysine 9 (H3K9me) [19, 20]. The assembly of heterochromatic domains requires the joint action of a series of chromatin-modifying enzymes [12, 20].

In *Drosophila* and mammals, 12 factors appear to form the key components of constitutive heterochromatin in somatic cells. These include: histone H1 and its H2a/z variant, the chromatin-modifying enzymes SUV39h1, SUV39h2, SUV4-20 h1, SUV4-20 h2, Hdac2, HPα, and HP1β, the group of high mobility proteins such as Hmga1 and Hmga2, remodeling components such as Atrx, Trim28 co-repressor, and members of the Mbd protein family (methyl-binding domain) [21].

Cytologically, facultative and constitutive heterochromatin regions were indistinguishable. Therefore, cytogenomic studies that use conventional C banding to detect heterochromatin lack the resolution required to determine which type of heterochromatin is contained in each genomic region. However, this technique remains important for demonstrating the considerable genomic plasticity shown by organisms in the face of different environmental stimuli. It is beyond the scope of this review to conduct a detailed survey of the composition of both types of heterochromatin, as our aim is to demonstrate how changes in the process of heterochromatin modulation, via retroelement inactivation or gene expression regulation, affect phenotypic plasticity of Amazon region fish when confronting different environmental stressors.

2.1 Epigenetic adaptation and environmental stress: focus on selected Amazonian fish

The link between epigenetic adaptation and response to environmental stimuli is a topic that has been studied extensively in recent years. Several studies have reported the emergence of epigenetic plasticity and the consequent assembly of heterochromatin in various organisms when exposed to stress [22–25]. Such studies have shown a close relationship between gene silencing, via heterochromatin assembly or DNA methylation, and transcriptional regulation.

In Amazonian fish species, a large number of heterochromatic patterns have been reported following chromosome C banding treatments. These have been related to a wide variety of environmental stressors.

In the loricariid catfish *Hypancitrus debilittera*, Silva et al. [26] identified several polymorphisms, including nucleolar organizing regions (RONs), 5S, and 18S rDNA ribosomal sites, mainly via C banding patterns. Likewise, an intense morphological staining polymorphism, designated as P1, was found in this species. In addition, pronounced morphological variation in the body striping pattern (termed P2, P3, P4, and P5) was recorded, suggesting a correlation with the observed chromosomal polymorphism (**Figure 2**). It was proposed that this morphological polymorphism

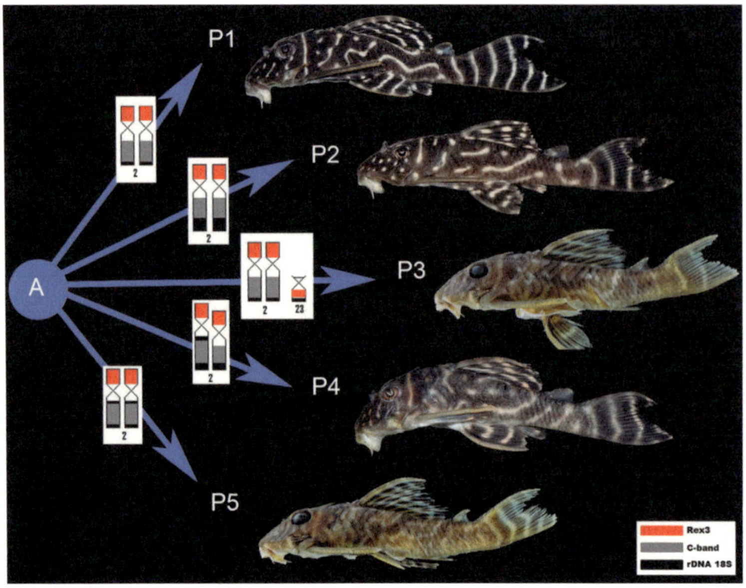

Figure 2.
Different morphological patterns (P1, P2, P3, P4, P5) and C-banding corresponding in Hypancistrus debillitera *highlighting the polymorphic pair 2, which presents a difference in heterochromatic block size and 18S rDNA sites. (more details see: [26]).*

was the result of a position effect variegation. Position effect variegation (PEV) was first described for the *Drosophila* gene *white,* and is a classic example of how constitutive heterochromatin operates, since, due to translocation events, this gene is variably silenced if located in a position adjacent to the pericentric heterochromatin [12]. The PEV phenomenon has been recorded widely in eukaryotes, from yeasts to humans [3, 21, 27], and can explain the observed variation in *Hypancitrus debilittera* body stripe patterns.

Heterochromatin-related PEV modifiers are called variegation suppressors [Su (var)], while those related to euchromatin are termed variegation enhancers [E (var)] [21]. Su (var) mutations weaken the formation and maintenance of heterochromatin, while E (var) mutations decrease euchromatin or allow heterochromatin expansion [5].

In *Hypancistrus debilittera,* C banding shows a variation in the distribution of heterochromatin in the five known different patterns (P1, P2, P3, P4, and P5), where heterochromatin assembly at gene regions, related to the staining pattern and consequent silencing, could result in the morphological polymorphism observed in the species. However, Silva et al. [26] reported no numerical or structural chromosomal rearrangements related to the emergence of polymorphic heterochromatin patterns. Therefore, to test the hypothesis of operative gene silencing by assembling PEV-like heterochromatin, the Rex3 transposable element was mapped and found to be present and associated with heterochromatic regions in multiple regions of the genome of this species [26]. This indicates that heterochromatin polymorphism may be associated with morphological polymorphism and other chromosomal polymorphisms found in the species (**Figure 2**).

Heterochromatin formation is strongly linked to transposable element inactivation [5]. Several studies have shown that depletion of Su (var) 3–9 (variegation-suppressing enzyme 3–9), which is a methyltransferase promoting H3K9 trimethylation [28, 29], can lead to the formation of mutant phenotypes, including abnormal chromosomal segregation, interruption of spermatogenesis (with links to hypogonadism and infertility), and increased risk of tumorigenesis [16]. In *Drosophila,* this protein is essential for the maintenance of nucleolar stability, and its loss promotes nucleolar fragmentation. In addition, aberrant recombination of repeated DNA sequences results in rDNA locus instability [9]. This suggests that the heterochromatin, rDNA, and Ag-RONs polymorphisms seen in *H. debilittera* are directly or indirectly related to the functioning of such enzymes as Suvar3–9 since the polymorphisms of rDNA and Ag-RONs can be indicative of loss or gain of protein functions active in the assembly of heterochromatin.

Hypancistrus debilittera (family Loricariidae) is endemic to the Xingu River [30], and preferentially inhabits rapids [31]. Structural chromosomal polymorphisms are common in the Loricariidae, with karyotype plasticity marked by Robertsonian rearrangements [26, 32–34]. Therefore, the polymorphisms found in this species can be explained as either reflecting the cellular mechanisms of an adaptive response to the environment of the rapids or may simply result from the genomic plasticity of the family.

Identification of the molecular signature of heterochromatin, under assemblages in the genome of *H. debilittera* that block the harmful action of transposable elements such as Rex3, may lead to an understanding of the mechanism producing such a high degree of polymorphism in these organisms. However, it remains unclear whether the color polymorphism of *H. debilittera* is a PEV-like phenomenon, related to an adaptation to environmental stimuli and the assembly of heterochromatin correlated with the inactivation of retrotransposable elements (for example Rex3), or whether it is an efficient mechanism for increasing genetic variability in the face of environmental challenges (such as those found in rapids). Answers to these questions are likely to involve interrelation between the two possibilities.

In a conceptually linked study, Whitelaw and Martin [35] analyzed isogenic strains of mice and found morphological phenotypic variation related to the action of retrotransposons. According to these authors, the effects of the stochastic activity of retroelements on gene expression and the inactivation process of these elements indicate that somatic cells of individuals can be epigenetic mosaics, corresponding to the activity of each retrotransposon, and such activity can produce subtle phenotypic variations, even in genetically identical individuals.

In another study, using the corydoras catfish *Hoplosternum littorale* (family Callichthyidae) as a model, Silva et al. [36] found polymorphism related to the presence of multiple 18S rDNA sites, co-located with Rex3 retroelements, from individuals from polluted and unpolluted forest streams (*igarapés*) bordering the Amazonian city of Manaus. They reported Rex3 sites to be more conspicuous in samples from polluted environments than those from unpolluted environments. In addition, the C band showed heterochromatin polymorphism, which is common for species of the fish family Callichthyidae, with samples from polluted environments having more conspicuous blocks (Igarapé Mindú and Igarapé Quarenta). It has been suggested that this heterochromatin increase results either from a heterochromatinization process or from the addition of heterochromatin, caused by uneven crossing-over, duplication, or epigenetic mechanisms such as DNA methylation and chromatin remodeling. In a complementary study, Silva et al. [37] compared cytogenetically polluted and unpolluted aquatic environments of Manaus, using three species of Amazonian fish, *Pterygoplichthys pardalis*, *Semaprochilodus insignis,* and *Cichlasoma amazonarum*. The authors found an increase in Rex6 in *P. pardalis*, in Rex1, and Rex3 in *S. insignis* and in Rex6 in *C. amazonarum*. There was a correlation between Rex6 and constitutive heterochromatin and increased heterochromatin in *P. pardalis* individuals from a polluted environment. In this work, the authors suggest the retrotransposable element Rex 6 as a marker of inhospitable environments.

Examining the relationship between retroelements and heterochromatin polymorphism, Silva et al. [36], found a significant increase in heterochromatin in *Colossoma macropomum* exposed to cupric sulfate at 30% of the species LC_{50}, compared to the group control. Changes in heterochromatin levels were evident after 48 h of stress, and after 72 h, several chromosomes appeared marked with heterochromatin. In addition, there were significantly more copies of the Rex1 retroelement in individuals exposed to $CuSO_4$ for 72 h, when compared to the control. This element was mapped by FISH in animals exposed for 48 h to $CuSO_4$, and several chromosomes appeared stained.

A study of the parental species and hybrid offspring (commonly known as *tambacu*) of crosses between female tambaqui (*Colossoma macropomum*) and male pacu (*Piaractus mesopotamicus*), Ribeiro et al. [38] found variation in the heterochromatin pattern of hybrids as well as conspicuous patches of Rex3 and Rex6 retrotransposable elements. This result was interpreted as arising from the need for adjustments in cell division and introgressive hybridization, since the joining of two different genomes frequently leads to changes in the control of gene expression, DNA methylation, chromosomal rearrangements, and, consequently, transposable element mobilization.

Ferreira et al. [39] exposed *Colossoma macropomum* to three climate change scenarios of the Intergovernmental Panel on Climate Change (IPCC). They reported chromosomal heterochromatinization in individuals exposed to the A2 climate scenario, in which terminal and interstitial bands of bands were observed in several chromosomes. These correlated with the presence of Rex3 retrotransposon elements. In another study involving tambaqui, Costa et al. [40] detected different patterns in heterochromatin distribution associated with increased Rex3 expression, following exposure to the parasiticide Trichlorfon. The authors found that,

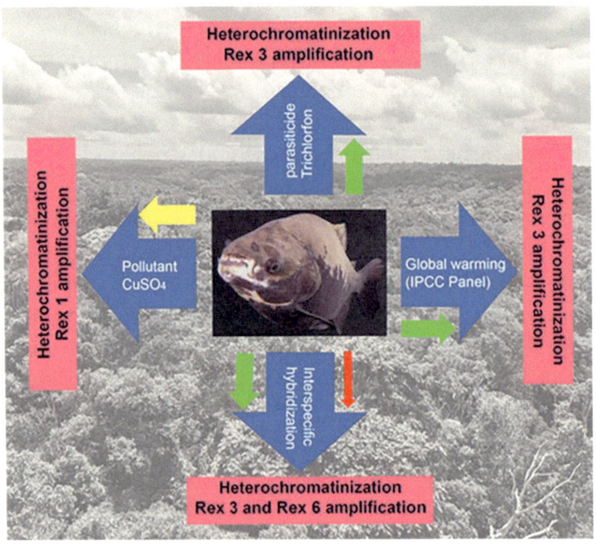

Figure 3.
The experimental data obtained in Colossoma macropomum *indicate that in all conditions of chemical stress, change in environmental temperature or shock between interspecific genomes, there was an increase in the amount of heterochromatin in the pericentromeric portion of all chromosomes of the complement or in the telomere portion, followed by an increase in the number of copies of the transposable elements. From left to right: Heterochromatinization and amplification of Rex1 in exposure to CuSO4 (yellow arrow); heterochromatinization and amplification of Rex3 in contact with antiparasitic Triclorfon and climate change provided for in the IPCC (green arrow); finally, amplification of Rex3 (green arrow) and Rex6 (red arrow) accompanied by heterochromatinization in a situation of interspecific artificial hybridization (more details see: [36–39]). Yellow arrow = Rex1, green arrow = Rex3 and red arrow = Rex6.*

after exposure to 50% of the established LC_{50}-96 h [41, 42] for the species, the organisms showed areas indicating active Rex3 elements as well as conspicuous heterochromatin marks in non-pericentromeric regions, with interstitial and terminal bi-telomeric markings. Such studies seem to show an intimate relationship between changes in the heterochromatic profile of *C. macropomum* and Rex3 retrotransposable element inactivation mechanisms.

The molecular composition of heterochromatin has not been elucidated in any of the aforementioned studies, although it has been correlated with Rex3 elements in *H. debilittera* [26] and *C. macropomum* [39, 40]. However, the role of retrotransposable elements in the regulation of gene activity seems undeniable. Mediated by heterochromatin, this produces widely variable phenotypes, which could help explain the enormous phenotypic plasticity observed in aquatic organisms in the Amazon. Future approaches focusing on methylation patterns in DNA and histone tails, in addition to detection of the molecular signature of heterochromatin assembled to inactivate both retroelements and other genes, should provide a clearer view of the adaptive processes developed by fish in the Amazon to deal with stressful environments.

In this context, tambaqui (*Colossoma macropomum*), a charismatic and often-studied Amazonian fish, is notable, as it shows the same responses to experimental simulation conditions, global warming conditions, treatment with antiparasiticides in cultivation, environmental pollution, and interspecific hybridization (**Figure 3**). It is not unreasonable to assume, under such circumstances, that the similarity of chromosomal behaviors under such a variety of stressors is casual and is likely to be the result of similar, if not identical, molecular and physiological mechanisms.

3. Conclusion

In conclusion, the species of Amazonian fish studied for heterochromatin assembly and retroelement dispersion (especially Rex 3) seem to respond dynamically and with remarkable similarity to a range of stressing stimuli.

Acknowledgements

The current study was supported in part by INCT ADAPTA II funded by CNPq – Brazilian National Research Council (465540/2014-7), FAPEAM – Amazonas State Research Foundation (062.1187/2017) and FAPEAM/SEPLANCTI/Governo do Estado do Amazonas - POSGRAD Res. No. 002/2016, and CAPES - Coordination for the Improvement of Higher Education Personnel. Programa de Pós-Graduação em Genética, Conservação e Biologia Evolutiva (INPA/GCBEv), Laboratório de Genética Animal (LGA), Universidade Federal do Amazonas (UFAM/LABICA), FAPEAM/SEPLANCTI/Governo do Estado do Amazonas - POSGRAD Res. N° 310 002/2016. Adrian Barnett helped with English.

Conflict of interest

The authors declare no conflict of interest.

Author details

Daniele Aparecida Matoso[1], Maelin da Silva[2], Hallana Cristina Menezes da Silva[3], Eliana Feldberg[4] and Roberto Ferreira Artoni[5*]

1 Department of Genetics, Institute of Biological Sciences, Federal University of Amazonas, Manaus, Brazil

2 Department of General Biology, State University of Ponta Grossa, Paraná, Brazil

3 Program Postgraduate in Genetics, Conservation and Evolutionary Biology, National Amazon Research Institute, Manaus, Brazil

4 National Amazon Research Institute, Manaus, Brazil

5 Department of Structural and Molecular Biology and Genetics, State University of Ponta Grossa, Ponta Grossa, Brazil

*Address all correspondence to: rfartoni@gmail.com

References

[1] Sumner AT. Chromosomes: Organization and function. Blackwell Science Ltda, 2003; 287p.

[2] Grewal SI, Jia S. Heterochromatin revisited. Nat Rev Genet. 2007; 8 (1):35-46. DOI: 10.1038/nrg2008. PMID: 17173056. DOI: 10.1038/nrg2008

[3] Lucchesi JC. Epigenetics, Nuclear Organization and Gene Function: With Implications of Epigenetic Regulation and Genetic Architecture for Human Development and Health. Oxford University Press; 2019. DOI: 10.1093/oso/9780198831204.001.0001

[4] Heitz E. Das heterochromatin der moose. I Jahrb Wiss Botanik. 1928;**69**:762-818

[5] Allis C, Jenuwein T, Reinberg D, Caparros M. Epigenetics. Vol.2, 2nd ed. New York: Cold Spring Harbor Laboratories Press, 2015. 984p.

[6] Biscotti MA, Olmo E, Heslop-Harrison JS. Repetitive DNA in eukaryotic genomes. Chromosome Research. 2015;**23**:415-420. DOI: 10.1007/S10577-015-9499-z

[7] Lyon MF. Gene action in the X-chromosomes of the mouse. Nature, 1961, 190:372-373.

[8] Wang J, Reddy BD, Jia S. Rapid epigenetic adaptation to uncontrolled heterochromatin spreading. eLife. 2015;**0**:1-17. DOI: 10.7554/eLife.06179

[9] Saksouk N, Simboeck E, Déjardin J. Constitutive heterochromatin formation and transcription in mammals. Epigenetics and Chromatin. 2015;**8**:1-17

[10] Quina AS, Buschbeck M, Di Croce L. Chromatin structure and epigenetics. Biochemical Pharmacology. 2006;**72**:1563-1569

[11] Hunter RG, McEwen BS, Pfaff DW. Environmental stress and transposon transcription in the mammalian brian. Mobile Genetic Elements, 2013; 3:e24555. DOI: 10.4161/mge.24555

[12] Wang J, Jia ST, Jia S. New insights into the regulation of heterochromatin. Trends in Genetics. 2016;**32**:284-294

[13] Trojer P, Reinberg D. Facultative heterochromatin: Is there a distinctive molecular signature. Molecular Cell. 2007;**28**:1-13. DOI: 10.1016/j.tig.2016.02.005.

[14] Harshman SW, Young NL, Parthun MR, Freitas MA. H1 histones: Current perspectives and challenges. Nucleic Acids Research. 2013;**41**:9593-9609. DOI: 10.1093/nar/gkt700

[15] Richards EJ, Elgin SCR. Epigenetic codes for heterochromatin formation and silencing: Rounding up the usual suspects. Cell. 2002;**108**:489-500

[16] Peters AHFM, Mermoud JE, O'Carroll D, Pagani M, Schweizer D, Brockdorff N, et al. Histone H3 lysine 9 methylation is an epigenetic imprint of facultative heterochromatin. Nature Genetics. 2002;**30**:77-80

[17] Jamieson K, Wiles ET, McNaught KJ, Sidoli S, Leggett N, Shao Y, et al. Loss of HP1 causes depletion of H3K27me3 from facultative heterochromatin and gain of H3K27me2 at constitutive heterochromatin. Genome Research. 2016;**26**:97-107. DOI: 10.1101/gr.194555.115

[18] Stunnenberg R, Shylini RK, Keller C, Kirschmann M, Gelman L, Buhler M. H3K9 methylation extends across natural boundaries of heterochromatin in the absence of an HP1 protein. The EMBO Journal. 2015;**34**:2789-2803. DOI: 10.15252/embj.201591320

[19] Allshire RC, Madhani HD. Ten principles of heterochromatin formation and function. Nature Reviews. Molecular Cell Biology. 2017;**19**(4):229-244. DOI: 10.1038/nrm.2017.119

[20] Noma KI, Allis CD, Grewal SIS. Transitions in distinct histone H3 methylation patterns at the heterochromatin domain boundaries. Science. 2001;**293**:1150-1155

[21] Fodor BD, Shukeir N, Reuter G, Jenuwein T. Mammalian Su (var) genes in chromatin control. Annual Review of Cell and Developmental Biology. 2010;**26**:471-501. DOI: 10.1146/annurev.cellbio.042308113225

[22] Ayoub N, Goldshmidt I, Cohen A. Position effect variegation at the mating-type locus of fission yeast: A cis-acting element inhibits covariegated expression of genes in the silent and expressed domains. Genetics. 1999;**152**:495-508

[23] Hunter RG, McCarthy KJ, Milne TA, Pfaff DW, McEwen BS. Regulation of hippocampal H3 histone methylation by acute and chronic stress. Proceedings of the National Academy of Sciences of the United States of America. 2009;**106**:20912-20917. DOI: 10.1073/pnas.0911143106

[24] Seong KH, Li D, Shimizu H, Nakamura R, Ishii S. Inheritance of stress-induced, ATF-2-dependent epigenetic change. Cell. 2011;**145**:1049-1061. DOI: 10.1016/j.cell.2011.05.029

[25] Andrés F, Coupland G. The genetic basis of flowering responses to seasonal cues. Nature Reviews Genetics. 2012;**13**:627-639. DOI: 10.1038/nrg3291

[26] DaSilva M, Ribeiro ED, Matoso DA, Sousa LM, Hrbek T, Py-Daniel LR, et al. Chromosomal polymorphism in two species of *Hypancistrus* (Siluriformes: Loricariidae): An integrative approach for understanding their biodiversity. Genetica. 2014;**142**:127-139. DOI: 10.1007/s10709-014-9760-y

[27] Grunstein M, Gasser SM. Epigenetics in *Saccharomyces cerevisiae*. Cold Spring Harbor Perspectives in Biology. 2013;**5**:1-28. DOI: 10.1101/cshperspect.a017491

[28] Schotta G, Ebert A, Krauss V, Fischer A, Hoffmann J, Rea S, et al. Central role of drosophila SU(VAR)3-9 in histone H3-K9 methylation and heterochromatic gene silencing. The EMBO Journal. 2002;**21**(5):1121-1131

[29] Schotta G, Ebert A, Dorn R, Reuter G. Position-effect variegation and the genetic dissection of chromatin regulation in drosophila. Seminars in Cell & Developmental Biology. 2003;**14**(1):67-75

[30] Rosa RS. Lima FCT. Os peixes brasileiros ameaçados de extinção. In: Machado BM, Drummond GM, Paglia AP, editors. Livro vermelho da fauna brasileira ameaçada de extinção. Brasília, MMA. 2008. pp. 9-285

[31] Gonçalves AP. Ecologia e etnoecologia de *Hypancistrus zebra* (Siluriformes: Loricariidae) no rio Xingu. Belém, PA: Amazônia brasileira. Dissertação de Mestrado. Universidade Federal do Pará; 2011 137p

[32] Giuliano-Caetano L. Polimorfismo cromossômico Robertsoniano em populações de *Rineloricaria latirostris*. Pisces: Loricariidae). PhD Thesis. Universidade Federal de São Carlos; 1998. p. 78

[33] Mendes-Neto EO, Vicari MR, Artoni RF, Moreira-Filho O. Description of karyotype in *Hypostomus regani* (Ihering, 1905) (Teleostei, Loricariidae) from the Piumhi river in Brazil with comments on karyotype variation found in *Hypostomus*. Comparative Cytogenetics. 2011;**5**:133-142. DOI: 10.3897/JHR.v5i2.964

[34] Ziemniczak K, Barros AV, Rosa KO, Nogaroto V, Almeida MC, Cestari MM, et al. Comparative cytogenetics of Loricariidae (Actinopterygii: Siluriformes): Emphasis in Neoplecostominae and Hypoptopomatinae. Italian Journal of Zoology. 2012;**79**:492-501. DOI: 10.1080/11250003.2012.676677

[35] Whitelaw E, Martin DIK. Retrotransposons as epigenetic mediators of phenotypic variation in mammals. Nature Genetics. 2001;**27**:361-365

[36] Silva FA, Schneider CH, Feldberg E, Baccaro FB, Carvalho NDM, Gross MC. Genomic organization under different environmental conditions: *Hoplosternum littorale* as a model. Zebrafish. 2016;**13**(3):197-208. DOI: 10.1089/zeb.2015.1237

[37] Silva FA, Guimarães EMC, Carvalho NDM, Ferreira AMV, Schneider CH, Carvalho-Zilse GA, et al. Transposable DNA elements in Amazonian fish: From genome enlargement to genetic adaptation to stressful environments. Cytogenetic and Genome Research. 2020;**160**(3):148-155. DOI: 10.1159/000507104

[38] Ribeiro LB, Moraes-Neto A, Artoni RF, Matoso DA, Feldberg E. Chromosomal mapping of repetitive sequences (Rex3, Rex6, and rDNA genes) in hybrids between *Colossoma macropomum* (Cuvier, 1818) and *Piaractus mesopotamicus* (Holmberg, 1887). Zebrafish. 2017;**14**(2):155-160. DOI: 10.1089/zeb.2016.1378

[39] Ferreira AMV, Marajó L, Matoso DA, Ribeiro LB, Feldberg E. Chromosomal mapping of rex retrotransposons in Tambaqui (*Colossoma macropomum* Cuvier, 1818) exposed to three climate change scenarios. Cytogenetic and Genome Research. 2019;**159**:39-47. DOI: 10.1159/000502926

[40] Costa MS, Silva HCM, Soares SC, Favarato RM, Feldberg E, Gomes ALS, et al. Distribution of Heterocromatin and Retrotransposable Elements *Rex* 1, 3 and 6 in Tambaqui (*Colossoma macropomum*) in Captivity Exposed to the Parasiticide Trichlorfon. Manaus, Amazonas: Dissertação de Mestrado; 2020 40p

[41] Silva HCM, da Silva AG, Idalino JJS, Sousa FB, Gomes ALS, Duncan WP, et al. Trichlorfon acute lethal toxicity to juvenile tambaqui (*Colossoma macropomum*). Aquaculture Research. 2020;**51**:863-866. DOI: 10.1111/are.14412

[42] Duncan WP, Idalino JJS, da Silva AG, Moda RF, da Silva HCM, Matoso DA, et al. Acute toxicity of the pesticide trichlorfon and inhibition of acetylcholinesterase in *Colossoma macropomum* (Characiformes: Serrasalmidae). Aquaculture International. 2020;**28**:815-830. DOI: 10.1007/s10499-019-00497-w

Chapter 4

Chromosome Banding and Mechanism of Chromosome Aberrations

Sanjay Kumar, Asikho Kiso and N. Abenthung Kithan

Abstract

Chromosome identification depends on the morphological features of the chromosome and therefore karyotype and its banding pattern analyses are the most suitable technique to identify each and every chromosome of a chromosome complement. Moreover, aberrations caused by breaks play an important role in the evolution of a chromosome set and chromosome complement by decreasing or increasing the chromosome number. Therefore, both the aspects are discussed in detail in the present chapter. At present, the chapter will highlight the karyotype and its components, karyotype trends, evolution and its role in speciation, banding pattern and techniques, chromosome differentiation and linearization, banding applications and their uses, detection and analysis of chromosomal aberrations, chromosome and chromatid types of aberrations and mechanism of the formation of chromosome aberrations and breaks for karyotype evolutionary trends.

Keywords: chromatin, karyotypes, karyoype trend, karyotype evolution, chromosome banding techniques and pattern, chromosome aberrations and detection

1. Introduction

Chromatin is a mixture of DNA, RNA and proteins could be easily visible during the interphase and prophase of the cell division cycle. Chromatin from interphase (loose mixture of DNA, RNA Protein) to thick mitotic structure (tightly packed or compressed mixture of DNA, RNA Protein) packed through Nucleosome model of DNA packing. Chromatin is divided into euchromatin and heterochromatin. Euchromatin is a part of chromatin which takes less stain, loosely packed, genetically active, involved in active transcription, dispersed appearance with more DNA content than RNA. On the other hand, heterochromatin is slightly opposite to the euchromatin with dark stained region, tightly packed, genetically inactive, not involved in the active transcription, thick appearance with more RNA content than DNA. Heterochromatin could be of two type's constitutive and facultative heterochromatin. Constitutive heterochromatins are permanently conserved or condensed and in stable form i.e. not changed from heterochromatin to euchromatin and vice-versa. It consists of multiple repeats of DNA sequences with quite less density of genes in this region which are transcriptionally inactive. Most probably, thick and condensed state of the constitutive heterochromatin, replicates late in S-phase with reduced frequency of genetic recombination.

Facultative heterochromatins are not permanently conserved or condensed and in unstable form i.e. easily changed from euchromatin to heterochromatin and vice-versa [1–3].

Heterochromatic regions could be easily recognized on chromosome structure in the form of chromomeres, chromocentres and knobs. Chromomeres are regular features of all prophase chromosomes but their number, size, distribution and arrangements are specific for a particular species at a particular stage of development. Chromocentres are the regions with varying size near the centromere in the proximal regions of chromosome arms. Some of the Chromocentres could be resolved into large number of strings of chromomeres which are much larger in size as compared to the chromomeres found in the distal region of the chromosmes arms during the mid-prophase. The relative distribution of the chromocentres on the chromosome structure, sometimes considered to be of significant evolutionary value. Knobs are considered to be a spherical bodies or regions with spherical in shape and sometimes diameter of these spherical bodies is equal in width to chromosome arm, but the size may vary i.e. less or more than the diameter of chromosome arm. For example, a very distinct such type of chromosome knob could be observed in maize (*Zea mays*) at pachytene stage of meiosis I. It could be considered as a valuable chromosome marker for distinguishing chromosome of related species and races [4–6].

2. Concept of karyotype and its components

Karyotype may be defined as the study of chromosome morphology of a chromosome complement in the form of size, shape, position of primary constriction or centromere, secondary constriction, satellite, definite individuality of the somatic chromosomes and any other additional features. Karyotype highlights closely or distantly related species based on the similarity or dissimilarity of the karyotypes. For example, a group of species resemble each other in the number, size and form of their chromosomes. There may be 12 different types of karyotype categories depending on increasing asymmetry in chromosome complement [7]. The degree of asymmetry of chromosome complement depends on the four arm ratios (1 to 4) and the size of the smallest and largest chromosome and three different proportions of the metacentric chromosomes (ABC) of a given chromosome complement. Arm ratio 1 being the most symmetrical and 4 is the most asymmetrical. There are various quantitative karyotypic ratios to observe the karyotype variations and precise description of the karyotype such as relative length, centromeric index, total form percent, disprsin index, disparity index, coefficient of variation, volume of chromosomes, value of relative chromatin and so on. Asymmetric karyotype may be defined as the huge difference between the largest and smallest chromosome as well as less number of metacentric chromosomes in a chromosome complement. Similarly, symmetric karyotype may be defined as the small difference between the largest and smallest chromosome as well as more number of metacentric chromosomes in a chromosome complement [8, 9].

The principle ways in which karyotypes differ from each other are (i) basic chromosome number, (ii) form and relative size (V→J or L→I) of different chromosomes of the same set, (iii) number and size of satellites (related to those positions of the chromosome which form nucleoli) and secondary constrictions (NOR region of chromoosmes), (iv) absolute size of the chromosomes, (v) distribution of material with different staining properties i.e. euchromatin and heterochromatin [10, 11].

2.1 Karyotype trend

Karyotypic trend may be defined as the evolutionary changes in the chromosome complement by increasing or decreasing its base chromosome number which showed a definite direction of movement or pattern of its movement either from polyploidy to diploid or vice versa. For example, *Luzula* species (Juncaceae), also called wood rush, a monocot with holocentric chromosomes showed huge variation in genome and pattern or direction of chromosome movement from diploid to polyploidy or vice versa through symploidy and agmatoploidy phenomenon. The phenomenon could be related to the ascending or descending dysploidy which is also known as pseudoaneuploidy where chromosomes rearrange themselves within or between the chromosomes to decrease or increase the chromosome number in the chromosome complement of a particular species. Simploidy is the phenomenon of fusion of chromosomes together to reduce the chromosome number while the agmatoploidy breaks the chromosomes (fission) to increase the chromosome number for a particular species (**Figure 1**). The trend of *Luzula* species are as follows, L. purpureo-splendens (2n = 2x = 6; chromosome length 6.66 μm), L.elegans (2n = 2x = 6; chromosome length 4.62 μm), *L. alpinopilosa* (2n = 2x = 12 ± 1; chromosome length 2.55 μm), L.nivea (2n = 2x = 12; chromosome length 1.70 μm), L. sylvetica (2n = 2x = 12; chromosome length 1.48 μm), *L. multiflora* (2n = 6x = 36; chromosome length 1.32 μm), and *L. sudetica* (2n = 8x = 48; chromosome length 0.52 μm). The trend could be explained to understand that species with chromosome number 12 has merged their chromosomes together through a process of symploidy to occur speciation of a new diploid with 2n = 6. This could be possible because the size of the chromosomes increasing in L elegans and L. purpuro-splendans. Similarly, there is a possibility of agmatoploidy phenomenon has been occurred and the size the chromosomes decreased in *L. sudetica*. Moreover, it clearly

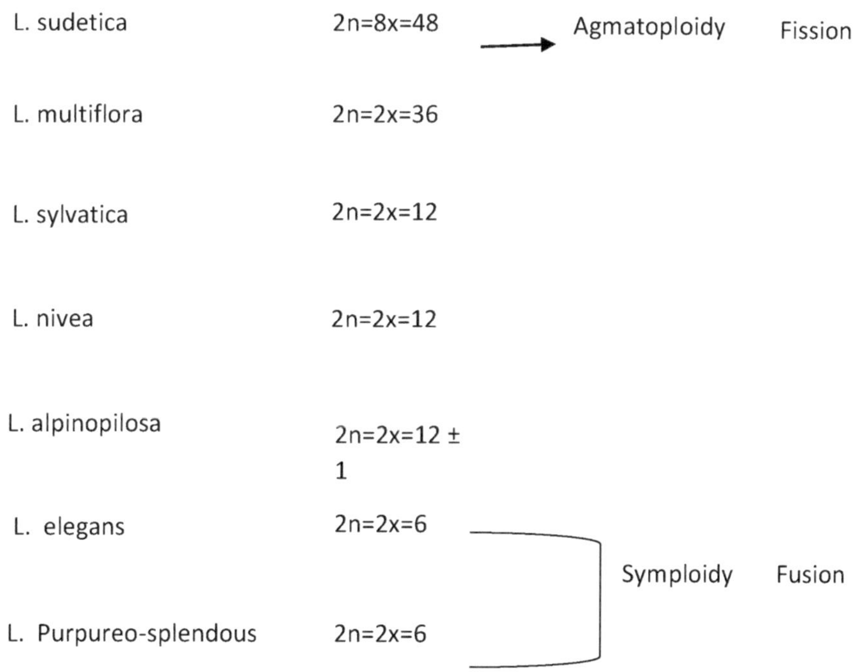

Figure 1.
Karyotype trend in Luzula *species by fission and fusion of the chromosomes.*

suggests the trend of chromosome size decrease from diploid species towards polyploidy species or vice-versa [12, 13].

2.2 Karyotype evolution and speciation

Karyotype evolution may be defined as a phenomenon of change in chromosome number with time and space where fusion or fission and rearrangement may take place among chromosomes to decrease or increase its chromosome complement as well as to adapt themselves in available climatic conditions at that particular place and for their survival over a period of time (**Figure 2**). For example, triticeae genome with a basic chromosome number 7 had undergone 5 centric and 7 nested fusion to reach the present 5 chromosome structure karyotype in *Brachypodium distachyon*. The fusion in the genome of triticeae and *B. distachyon* involved different combinations of ancestral chromosomes and therefore they were independent of each other [14]. When triticeae genome was crossed with *B. distachyon* genome [triticeae, 2n = 2x = 14 (TT) × *B. distachyon*, 2n = 2x = 10 (BB) = hybrid, 2n = 2x = 12 (TB)], a hybrid of 2n = 2x = 12 chromosomes was considered to be the ancestor of present day *B. distachyon* (2n = 2x = 10). It was considered that approximately 7 nested fusion (large number of breaks in the chromosomes and then repositioning of the fragments) suggest descending dysploidy in the ancestor to reach the present *B. distachyon* (2n = 2x = 10). The evolution of eudicot and monocot lineages is driven by two counteracting processes i.e. whole genome duplication (WGD) and diplodization. It is inferred that allthe present grass genomes evolved from an intermediate ancestor with 12 chromosomes which itself arose from 5 or 7 chromosome ancestor through WGD and two reciprocal translocations (**Figure 3**). Although this particular rearrangements is common in grasses, it rarely occurs in eudicots in which end to end fusion are mostly responsible for reduction in chromosome number [15, 16].

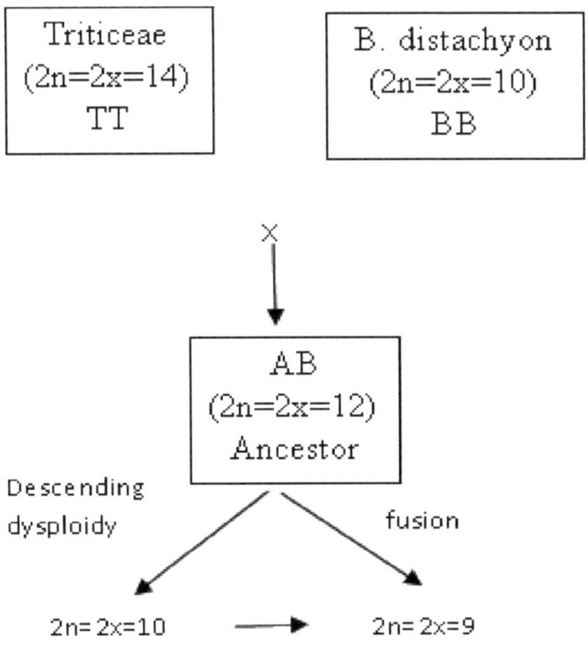

Figure 2.
Karyotype evolution in Brachypodium *by fusion or descending dysploidy from the ancestor.*

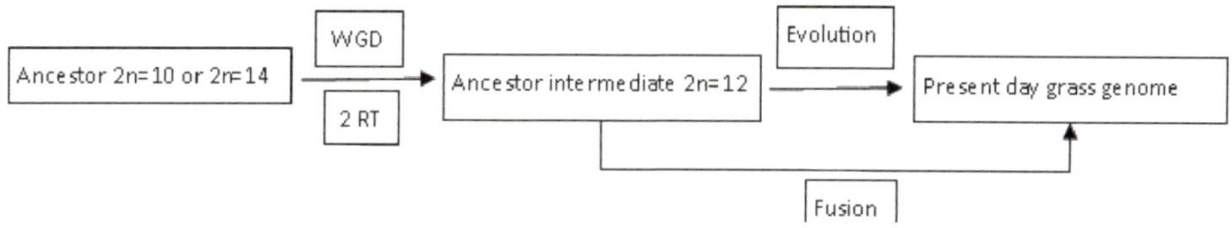

Figure 3.
Schematic diagram to show the karyotype evolution in grasses.

Similarly, evolutionary history of a karyotype is oftenly difficult to trace for older events and with time the accumulation of chromosome rearrangements remove the exact identity, number and order of the events occurred along the lineages leads to an extant karyotype. There are techniques to reconstruct the Extant karyotypes by extracting the information's from the extinct and its close relatives to get the hint for the direction of evolution of karyotype under paleogenomics [17].

3. Banding techniques

This is a technique for the identification of chromosomes and its structural abnormalities in the chromosome complement. Chromosome identification depends on their morphological characteristics such as relative length, arm ratio, presence and absence of secondary constrictions on the chromosome arms. Therefore, it is an additional and useful tool for the identification of individual chromosome within the chromosome complement. Further, it could be used for identification of chromosome segments that predominantly consist of either GC or AT rich regions or constitutive heterochromatin. The technique which involves denaturation of DNA followed by slow renaturation permits identification of constitutive heterochromatin as it mainly consists of repetitive DNA. On banded chromosome, darkly stained or brightly fluorescent transverse bands (positive bands) alternate with the lightly stained or less fluorescent (negative bands). The bands are consistent, reproducible and are specific for each species and each pair of homologous chromosomes. Banding techniques also revealed the extensive genetic polymorphism manifested as inter-individual differences in the size and stain ability of certain chromosomal segments. Initially four basic types of banding techniques were recognized for the identification of Human chromosome complement (Q, C, G and R bands) and later on two additional major type of bands were developed (N and T bands) for complete identification of the chromosome complement (**Figure 4**). Now present bands and newly developed bands or molecular

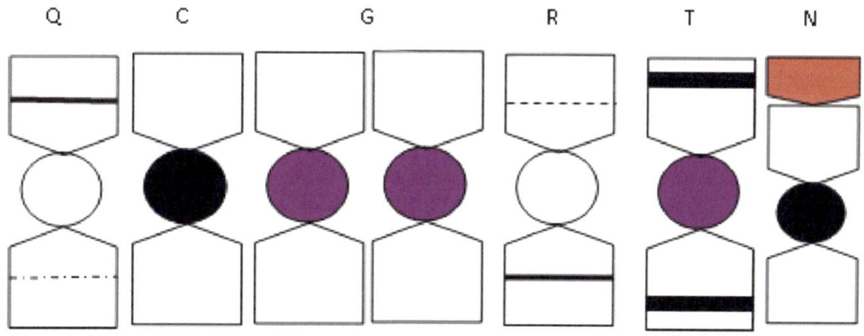

Figure 4.
Characteristics of various basic banding techniques.

bands are widely used in animals and plants for the identification of chromosome complement, chromosome aberrations as well as traces of phylogeny [18, 19].

3.1 Banding pattern of Q, G, R and C bands on Human chromosome complement

Chromosome band C and G clearly identifies the secondary constrictions of chromosome number 1, 9 and 16 sometimes slight or occasional staining were found for secondary constrictions of chromosome 9. C-band clearly stains and identifies peri-centromeric region on the chromosomes, while band Q slightly stains peri-centromeric region of chromosome 3. Both C and Q bands are equally important for staining the distal part of long arm of Y chromosome but for both the bands partial staining was recorded for satellites. Partial Q band staining was reported for chromosome 3, 13 and 21 while other chromosomes were recorded with intense staining. The C-band was found suitable to stain important regions and structures of the Human chromosome complement and widely used band. The G band is also known as golden band for the identification of the homologous pair within complement and could be considered a basic band before application of any sophisticated and molecular approach for further investigation [20].

3.2 Code system for banding pattern

There were 3 letter coding system for the banding procedure, for example, first letter codes for the type of banding to be done; second letter codes for the general technique to be used and third letter codes for the stain to be used. For instance, code QFQ indicates the Q-band to be done, fluorescence technique to be used and quinacrin mustard stain to be used during banding procedure. Similarly, other codes may be QFH, QFA, GTG, GTL, GAG, CBG, RFA, RHG, RBG, RBA, THG and THA depending on the bands, techniques and stains [21].

4. Chromosome bands

4.1 Q (quinacrine) band

The band stains the chromosome with fluorochromequinacrine mustard or quinacrinedihydrochloride (atebrin), observed under fluorescence microscope, and shows a specific banding pattern [22]. The fluorescence intensity is determined by the distribution of DNA bases along the chromosomal DNA with which the dyes interacts. The AT-rich regions enhance the fluorescence while GC-rich regions quench the fluorescence. The brightly fluorescent Q bands show high degree of genetic polymorphism but the fluorescence of Q band is not permanent and fades rapidly, therefore, the banding must be observed on fresh preparation and selected metaphases photographed immediately for further analysis. The disadvantage of the technique is the application of an expensive fluorescent microscope.

Q banding could also be achieved by fluorochromes other than quinacrine or its derivatives e.g. daunomycin, hoechst33258, BrdUetc which enhances AT-rich regions and quenches GC-rich regions. Acridineornage stains AT-rich regions red and GC rich regions green.

4.2 C (constitutive heterochromatin) banding

C banding was developed as a by product of in situ hybridization experiments on the localization of the mouse satellite DNA [23]. Centromeric regions with constitutive hterochromatin where satellite DNA was located stained more deeply with Geimsa than the rest of the chromosome [24]. The C banding technique is based on the denaturation and renaturation of DNA and the regions containing constitutive heterochromatin stain dark (C band) and could be visible near the centromere of each chromosome. The C bands are polymorphic in size which is believed to correspond to the content of the satellite DNA in those regions. C banding allows precise analysis of abnormalities in the centromeric regions and detection of isochromosomes. The C banding in combination with simultaneous T-banding in particular, extends to easy detection of dicentric rings [25]. Sometimes, C banding also permits to ascertain the parental origin of foetal chromosomes and distinguish between maternal and foetal cells in amniotic fluid cell culture.

4.3 G (Giemsa) banding

The banding could also be recognised as the modification of C banding procedure [26]. The technique permits the accurate identification of each pair of the chromosomal complement as well as recognition of the specific chromosomal rearrangements within complement. The preparations are permanent after staining with giemsa. A number of modifications for G bands have been developed and proposed such as pre-treatment with trypsin, urea, enzymes and salts, even though original ASG method (G like bands) as well as trypsin method often slightly modified and most widely used.

G bands correspond exactly to chromomeres of meiotic chromosomes but the mechanism leading to the visualization of the basic chromosome pattern is still unclear. The process is believed to be associated with denaturation and distributin of non-histone proteins and rearrangements of chromatin fibres from G negative to G positive bands.

4.4 R (Reverse) banding

R banding patterns are based on the thermal treatment of chromosomes and in general the reverse of the Q and G bands developed and proposed by Dutrillaux and Lejeune [27]. The ends of the R banded chromosomes are almost or always found positive and the centromeric regions are easily distinguished. This permits the observation of minor abnormalities in the terminal regions of chromosomes and the precise determination of chromosomal lengths. The technique is performed on a fixed chromosomal preparation and is based on heat denaturation of chromosomal DNA. R bands (GC rich regions) are more sensitive to DNA denaturation than Q and G bands (AT-rich regions). Giemsa stained R bands can be observed under phase contrast microscope while acridine orange stained R bands require fluorescence microscope.

4.5 T (Telomeric) banding

T bands are, in fact, the segments of the R bands that are most resistant to the heat treatment and the transition patterns between the R and T bands could be obtained by gradual treatments. Therefore, it may be regarded as the modifications of the R banding technique [28]. The clear marking of telomeric regions of chromosome with T banding enables the detailed analyses of the structural rearrangements at the ends of chromosomes. It also allows the detection of human chromosome 22 and its involvement in translocation. The usefulness of this method is for the detection of dicentric rings that were undetectable by other procedures. T bands can be observed either after giemsa or acridine orange staining.

4.6 N (Nucleolar organizing regions) banding

The NOR regions could be selectively stained by techniques involving either giemsa or silver staining. The giemsa technique developed by Matsui and Sasaki [29] allows the staining NOR after extraction of nucleic acids and histones. The technique N banding was improved by Funaki et al. [30]. The silver staining technique fall into two categories; a) Ag-As method: the method is based on the staining with combined silver nitrate and ammoniacal silver solutions; b) Ag method: in this method, staining with ammoniacal silver is omitted. NOR banding stains only those regions that were active as nucleolus organizers in the preceding interphase as well as useful for visualization and study of satellite associations.

4.7 Sequential banding

In routine cytogenetic diagnosis, a single banding technique is usually sufficient for the detection of chromosomal abnormalities e.g. G banding or R banding, but sometimes, more complicated chromosomal rearrangements often require sequential staining of the same metaphase by several banding techniques and the process is known as sequential banding. The quality of chromosomes in sequential banding deteriorates with each staining therefore; it restricts the sequential banding up to 3 or 4 different staining techniques (**Figure 5**). For example, single metaphase ➔ First procedure, Q banding ➔ Second procedure, G banding ➔ Third procedure, C banding ➔ deteriorates the chromosome quality ➔ therefore, restricts up to 3 or 4 staining procedures [31].

4.8 Simultaneous banding

This is the technique that produces simultaneously two types of banding on the same metaphase or on one slide but different metaphases (**Figure 5**). For example, same metaphase ➔ first procedure, G banding ➔ second procedure, C banding OR single slide with different metaphases ➔ first procedure, C banding ➔ second procedure, T banding. Simultaneous banding restricts up to two staining procedures in different or single metaphase and results in precise estimation of chromosomal aberrations [32].

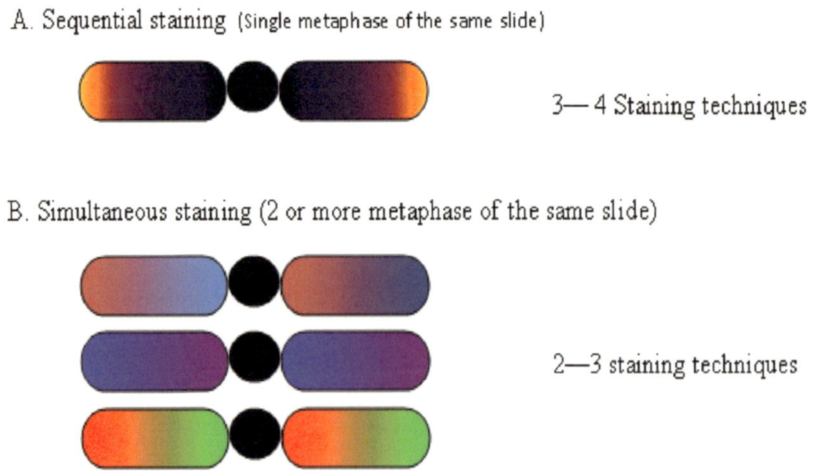

Figure 5.
Banding techniques for sequential and simultaneous staining.

4.9 High resolution banding

The general trend is to pick highly thick and contracted metaphase chromosomes are chosen for banding procedures which gives enough banding patterns for prediction of any kind of aberrations or arrangements within complement, but sometimes it fails also. Therefore, elongated chromosomes (earlier stages of mitotic divisions before reaching metaphase stage) are standardized for banding patterns and prediction of aberrations or arrangements called high resolution banding. Elongated chromosome standardization and preparation could be obtained through synchronization of cell cycle or use of various chromosome anti-contraction reagents through a procedure known as cell cycle synchronization technique. Cell synchronization is a process by which mitotic cells at different stages of cell cycle in culture are brought to the same phase through physical fractionation or chemical blockage or inhibition of DNA synthesis during S-phase. Cell synchrony may be defined as the progression of cells through cell cycle. The possible procedure for high resolution banding pattern are described. Mitotic stages (cultured cells) ➔ cell synchronization by adding chemical reagents ➔ Amethopterin or methotrexate or thymidine or fluorouracil ➔ cultured cells ➔ Blocks DNA synthesis in cultured cells ➔accumulation of cells in S-phase of cell cycle ➔ block released ➔ cell synchrony (large quantities of cells continue their cycle from the same level) ➔ prophase to mid-metaphase range ➔ high number of bands (gives more information as compared to the compact metaphase banding) [33, 34].

The technique allows precise localization of break points in chromosomal rearrangements and detection of minute chromosomal alterations that are undetectable by the mid-metaphase banding techniques [35].

5. Chromosome differentiation

Conventional chromosome banding techniques help us in understanding the patterns associated with chromosomal evolution of a species, speciation processes (formation of a new species) and generation of genetic variability among the species. Similarly, molecular banding techniques such as FISH and GISH provide little more and specific information by causing more differentiation in banding pattern of a chromosome of a particular species. Now, computational analyses (software packages) of the chromosome bands provide maximum information on banding pattern by increasing number of bands which helps to predict the specific and precise result on chromosomal aberrations and arrangements (**Figure 6**). For example, conventional chromosome bands ➔ enough bands for analysis ➔ molecular banding techniques ➔ more bands, more differentiation, more information ➔ computational techniques (software packages) ➔ still more bands, still more differentiation, still more information [36].

6. Chromosome linearization

Chromosome linearization is an important tool under computational analyses of the chromosome bands which suggest that linear chromosomes will provide maximum number of bands as well as information regarding aberrations or arrangements as compared to the twisted, rounded, curved and overlapped chromosomes. The information obtained from the straight chromosomes will be larger or maximum in quantity and quality. The tool of image linearization enables a better visualization technique which ultimately extends and refines the information that can be extracted from the chromosomes (**Figure 6**). For example, conventional chromosome bands ➔ enough bands for analysis ➔ molecular banding techniques ➔ more bands, more differentiation, more information ➔ computational techniques (software packages) ➔ better organization of cytogenetic data ➔ tool of image linearization ➔ still more bands, still more differentiation, still more information [37].

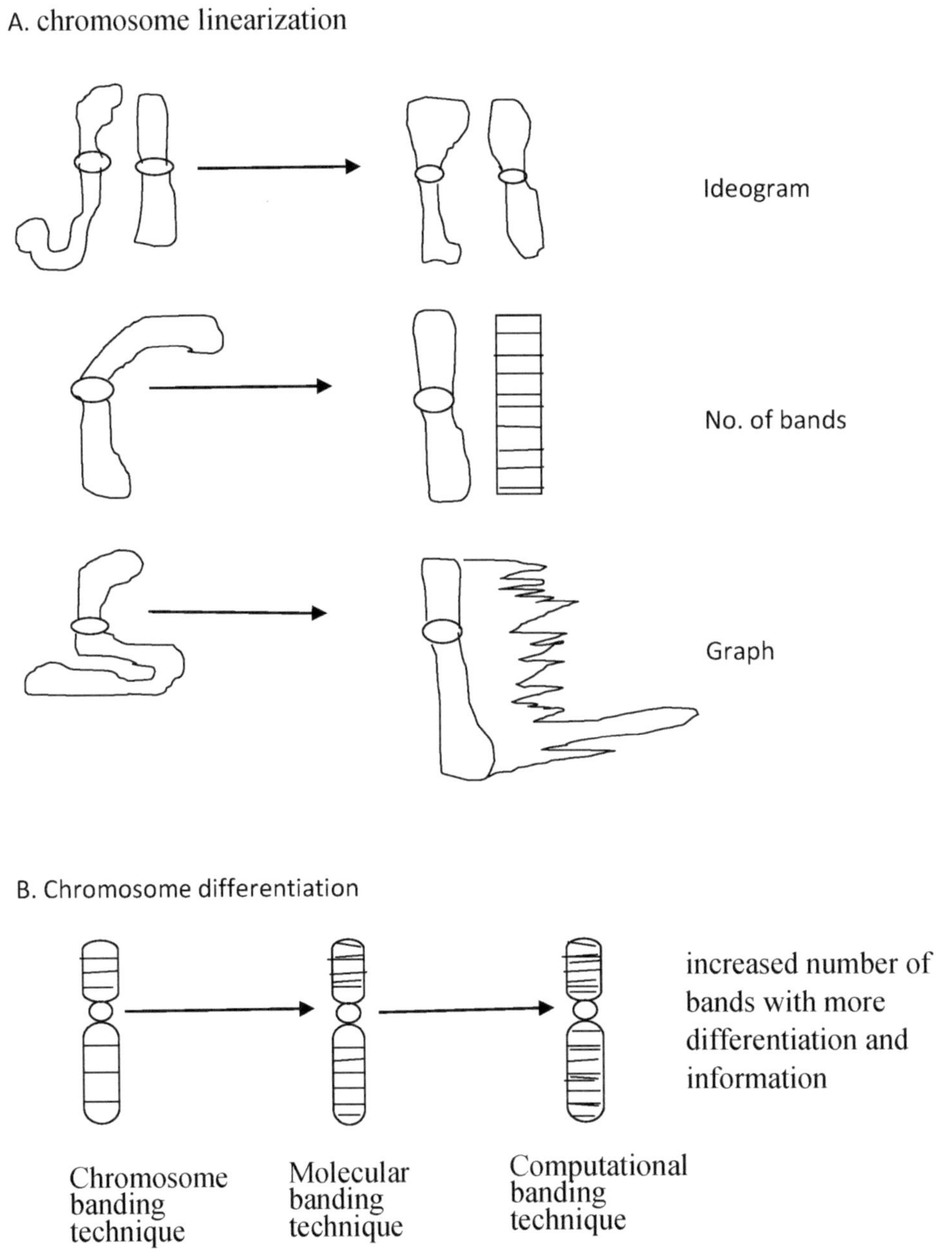

Figure 6.
Characteristics of chromosome linearization and differentiation banding technique.

7. Chromosome banding application

The chromosome banding primarily could be used for the detection and recognition of nature and type of the aberrations, identification of the chromosome involved and most importantly, the location of the presumptive positions of the lesions (usually termed as break points) involved in the structural changes in chromosome complements [38].

The basic requirement for the detection of chromosomal structural changes using the banding methods is disruption in the normal sequential band pattern of a chromosome arm region. The disruption in the chromosomal arm may take several forms and the most common forms are presence of an additional band, absence of an expected band, unexpected change in banding pattern and reversion of a part of banding pattern. The basic requirements for the chromosomal structural changes are possible by the existence of consistent and fixed pattern of chromosomal bands on chromosomal arm region as well as existence of good sequential differentiation between the chromosomal arm bands. Chromosome condensation is a process that occurs as the cell progresses towards metaphase and continues till the cells are held at metaphase by the action of spindle inhibiting chemicals. Consequently, number of bands, banding pattern, clarity and differentiation among the bands would be a function of state of chromosome contraction and possible provides the existence of consistent and fixed pattern of chromosomal bands at metaphase on chromosomal arm region. The good sequential differentiation between the chromosomal arm bands may cause differences in size, staining intensity, longitudinal spacing, numerical concentration of dark and light bands which forms the basis for the identification of individual chromosomes, but sometimes, the quality of differentiation varies between treatments and also between the cells on the same slide by application of different staining and banding procedures [39].

8. Detection and analysis of preliminary chromosomal aberrations

The primary chromosome aberrations could be detected and analysed in three forms such as achromatic lesions, chromosome type and chromatid type structural aberrations. The detection and characterization of aberrations or breaks which falls in the light banded region or euchromatic heterochromatin region which generally takes light stain could be identified as achromatic lesion. Chromosome type structural aberrations could be detected and identified in the form of asymmetrical interchanges (two translocations in the same arm), symmetrical interchanges (reciprocal translocations), inter arm intra changes (centric ring, pericentric inversions), intra arm intra changes (interstitial deletions, paracentric inversions), and breaks [(i) fragment compound (usually contains two terminal regions) and suspects an incomplete dicentric or centric ring; compound fragment with a terminal and unrelated non-terminal region suspects an incomplete complex interchanges, (ii) fragment simple (genetic material from one arm only) but with altered banding sequencing, suggest that sequence alteration is most frequently cause the inversion of a chromosomal proximal segment which is most probably indicates an incomplete paracentric inversion, (iii) fragment simple but the normal band sequence; (a) short arm with normal band sequences, if fragment present and not related to short arm might be incomplete reciprocal translocation or pericentric inversion; if fragment present and related to short arm may be true terminal deletion; (b) short arm with abnormal band sequences, first, observe for a distal inversion or incomplete paracentric inversion, second, observe for distal genetic material missing or for incomplete interstitial deletions, third, observe for presence of any minutes and fourth, if any extra additional distal segment is present but not related to short arm with abnormal band sequences, it may be regarded as incomplete complex changes and look for the origin of additional segment, (iv) fragment simple but non terminal; perhaps represents incomplete acentric ring or large interstitial deletion]. Chromatid type structural aberrations could be detected and identified in the form of interchanges (involve the interchanges of chromatids), interarm intra changes (requires two breaks and causes pericentric inversions, double duplications-deletions, dicentric and centric rings), intra arm intra changes (two breaks involved and caused the formation of isochromatid deletion, duplication-deletion, chromatid minutes and paracentric inversions), and breaks [{causes chromatid terminal deletions and may be of various types; (i) a tandem duplication may be present in the complete chromatid, (ii) a tandem duplication may be present either in centric or acentric portion of the incomplete chromatid, (iii) the origin of acentric fragment may be of intercalary type and the sister chromatid has a normal pattern but it may show a bending opposite to the site of deletion, (iv) a paracentric inversion could be possible adjacent to the break in the acentric fragment but other sister chromatid may be normal, (v) a paracentric inversion could be possible adjacent to the break in the centric region or portion but other sister chromatid may be normal}].Additional chromatid type structural aberrations could be possible in the form of as isochromatid (isochromatid exchanges), insertions and additional dark bands in one of the chromatid at the exchange point. It is important to locate the position of aberration or break points at some stage in the production of breakage and rejoining of chromatid threads as this is the event that causes disruption and could be observed using different techniques. A detectable point of breakage is generally referred to as 'break point' but 'presumptive break point' would be more appropriate in light of consideration [40].

9. Mechanism of formation of chromosome aberration

DNA breaks (especially dsDNA) is a serious threat for cell when unrepaired or misrepaired, as they can result in genomic instability or later on may lead to chromosomal alterations and even cell death. The chromosomal aberrations formation is one of the major alteration formed during dsDNA breaks. Moreover, it has been reported that during each cell division approximately 5000 ssDNA breaks were generated per nucleus and approximately 1% of total ssDNA breaks converted in dsDNA breaks. There were two theories forwarded for explanation of how chromosomal aberrations and its formation take place i.e. breakage and reunion theory and exchange theory. The breakage and reunion theory explains that breaks in chromosome may rejoin and form the original structure through restitution and the exchange theory might lead to exchange the aberration by rejoining another type of breaks. dsDNA breaks could be repaired by possibly three pathways i.e. homologous recombination repairing (HRR) which restores the original sequences, non-homologous DNA end joining (NHEJ) which usually generates and repairs small alterations such as base pair substitution, insertions, deletions at break sites etc., and single strand annealing (SSA) which may lead to the formation of interstitial deletions. HRR (requires one break) and NHEJ (requires two breaks) and SSA are important pathways for repairing of dsDNA breaks with one or two breaks in eukaryotic cells as well as mammalian cells [41].

Chromosomal aberrations may be caused by various physical and chemical factors such as ionizing radiations, chemicals and spontaneous dsDNA breaks e.g. endogenous reactive oxygen species, topoisomerases and replication errors.

There are various methods for the detection of chromosomal aberrations such as cytogentic testing [{(a) chromosomal aberration test, (b) micronucleus test; (i) giemsa staining method, (ii) cytokinesis-blocked micronucleus assay method, and (iii) flow-cytometry micronucleus assay method, (c) karyotyping}], molecular cytogentic testing [(a) fluorescence in situhybridization (FISH), (b) microarray comparative genomic hybridization (mCGH)], and prenatal screening to detect foetal abnormalities [42].

The clinical symptoms of chromosomal aberrations includes, a) chromosomal aberrations and spontaneous aberrations (e.g. congenital malformations, heart and renal malformations), b) chromosomal aberrations and cancer (e.g. chronic myeloid leukemia), c) behaviour peculiarities associated with chromosomal aberrations (e. g. turner syndrome), d) changes in course of adolescence and fertility (e.g. premature aging), e) pattern of dysmorphic signs in chromosomal aberrations (e.g. down syndrome, cri-du-cat syndrome) and f) congenital malformations and chromosome aberrations (e.g. dandy-walker malformation, gastrointestinal malformations, CNS-spina bifida) [43].

10. Conclusion

At present, most of the studies are surrounded towards the human health and disease caused by the interaction of genetic and environmental factors. Sometimes it is difficult to understand the genetic constitution or mechanism of chronic diseases. Therefore, it is very important to understand the mechanism behind the abnormal cells either genetic or environmental or both to solve the problems completely. The identification of a particular abnormality at the initial stage is crucial and banding techniques conventional or molecular provide such an opportunity. Banding and chromosome aberrations are played an important key role in the assessment of various risks faced by the genetic constitution of eukaryotic cells. Therefore, it may continue further to assess the risk of various kinds of ailments, diseases, and geno-toxicity induced by the radiations, pharmaceuticals, environmental and synthetic chemicals.

Author details

Sanjay Kumar[1*], Asikho Kiso[2] and N. Abenthung Kithan[2]

1 Department of Botany, Banaras Hindu University, Varanasi, India

2 Department of Botany, Nagaland University, Nagaland, India

*Address all correspondence to: ksanjay79@gmail.com

References

[1] Sindon R R (2005). Molecular biology: DNA twists and flips. Nature 437: 1097-1098.

[2] Monday T (2010). Characterization of RNA content of chromatin. Genome Res 20: 899-907.

[3] Nicodemi M, Pombo A (2014). Models of chromosome structure. Curr opin Cell Biol 28: 90-95.

[4] Jian Y, Xu C, Guo Z, Wang S, Xu Y and Zou C (2017). Maize (*Zea mays* L.) genome size indicated by 180 bp knob abundance is associated with flowering time. Scientific Reports 7: 5954

[5] Liu B, Yip R K, Zhou Z (2012). Chromatin remodeling, DNA damage repair and aging. Curr Genomics 13: 533-547.

[6] Zheng C, Hayes J (2003). Structures and interactions of the core histones til domains. Biopolymers 68: 539-546.

[7] Stebbins G L (1971). Chromosomal evolution in higher plants. Edward Arnold, London, UK.

[8] Holmquist G P, Ashley T (2006). Chromosome organization and chromatin modification: influence on genome function and evolution. Cytogenet Genome Res 114: 96-125.

[9] Macgregor H (2012). Chromosome revisited. Chromosome Research 20: 911-924.

[10] Judd B (1998). Genes and Chromosomes: A puzzle in three dimensions. Genetics 150:1-9.

[11] Sheval E V, Prusove A N, Kireev I I, Fais D, Polyakov V Y (2002). Organization of higher level chromatin structures (chromomere, chromonema and chromatin block) examined during visible light-induced chromatin photo-stabilization. Cell Biol Int 26: 579-591.

[12] Bozek M, Leitch A R, Leitch I J, Drabkova L Z, Kuta E (2012). Chromosome and genome size variation in Luzula (Juncaceae), a genus with holocentric chromosomes. Botanical Journal of Linnean society 170 (4): 529-541.

[13] Battaglia E (1994). Nucleosome and nucleotype: a terminological criticism. caryologia 47: 193-197.

[14] Idziak D, Hazuka I, Poliwczak B, Wiszynska A, Wolny E, Hasterok R (2014) Insight into the Karyotype Evolution of *Brachypodium* Species Using Comparative Chromosome Barcoding. PLoS ONE 9(3): e93503.

[15] Leitch IJ, Leitch AR (2013) Genome size diversity and evolution in land plants. In: Leitch IJ, editor. Plant Genome Diversity. Wien: Springer-Verlag. pp. 307-322.

[16] Adams K L, Wendel J F (2005). Polyploidy and genome evolution in plants. Curr Opin Plant Biol 8: 135-141.

[17] Comai L (2005). The advantages and disadvantages of being polyploidy. Nat Rev Genet 6: 836-846.

[18] Stebbins G L (1940). The significance of polyploidy in plant evolution. The American Naturalist 74: 54-66.

[19] Khandelwal S (1990). Chromosome evolution in the genus *ophioglossum* L. Botanical Journal of the Linnean Society 102: 205-217.

[20] O'Connor C (2008). Karyotyping for chromosomal abnormalities. Nature Education 1(1):27

[21] Edger B A, Orr-Weaver T L (2001). Endoreplication cell cycle: more for less. Cell 105: 297-306.

[22] Caspersson T, Farber S, Foley G E, Kudynowski J, Modest E J, Simonsson E, Wagh U, Zech L (1968). Chemical differentiation along metaphase chromosomes. Exp Cell Res 49: 219-222.

[23] Sumner A T (1972). A simple technique for demonstrating centromeric heterochromatin. Exp Cell Res 75: 304-306.

[24] Pardue M L and Gall J G (1970). Chromosomal localization of mouse satellite DNA. Science 168: 1356-1358.

[25] Chamla Y and Ruffie M (1976). Production of C and T bands in human mitotic chromosomes after heat treatment. Human Genetics 34: 213-216.

[26] Arrighi F E and Hsu T C (1971). Localization of heterochromatin in human chromosomes. Cytogenetics 10:81-86.

[27] Dutrillaux B and Lejeune J (1971). A new technic of analysis of the human karyotype. C R Acad Hebd Seances Acad Sci D 272: 2638-2640.

[28] Saccone S, De Sario A, Della Valle G, Bernardi G (1992). The highest gene concentrations in the human genome are in T-bands of metaphase chromosomes. Proc Natl Acad Sci USA 89: 4913 4917.

[29] Matsui S and Sasaki M (1973). Differential staining of nucleolus organizers in mammalian chromosomes. Nature 246: 148-150.

[30] Funaki K, Matsui S and Sasaki M (1975). Location of nucleolar organizers in animal and plant chromosomes by means of an improved N-banding technique. Chromosoma (Berl.) 49: 357-370.

[31] Jiang J and Gill B S (1993). Sequential chromosome banding and in situ hybridization analysis. Genome 36: 792-795.

[32] Tucker J D, Christensen M L, Carrano A V (1988). Simultaneous identification and banding of human chromosome material in somatic cell hybrids. Cytogenet Cell Genet 48: 103-106.

[33] Carson H L (1970). Chromosome tracers of the origin of the species. Science 168: 1414-1418.

[34] Guerra M (2016). Agmatoploidy and symploidy: a critical review. Genetics and Molecular Biology 39: 492-496.

[35] Ronne M (1989). Chromosome preparation and high resolution banding techniques. A review. J Dairy Sci 72: 1363-1377.

[36] Hsu T C (1973). Longitudinal differentiation of chromosomes. Annual Review Genetics 7: 153-176.

[37] Liang C P and Garrard W T (1999). Targetted linearization of DNA in vivo. Methods 17: 95-103.

[38] Schubert I, Lysak M (2011). Interpretation of karyotype evolution should consider chromosome structural constraints. Trends in Genetics 27: 207-216.

[39] Daniely M, Barkai G, Goldman B, Aviram-Golring A (1999). Detection of numerical chromosome aberrations by comparative genomic hybridization. Prenatal diagnosis 19:100-104.

[40] Savage J R (1977). Application of chromosome banding techniques to the study of primary chromosome structural changes. J Med Genet 14: 362-370.

[41] Cannan W J and Pederson D S (2016). Mechanisms and consequences of double strand DNA break formation in chromatin. J Cell Physiol. 231 (1): 3-14.

[42] Chatterjee N and Walker G C (2017). Mechanism of DNA damage, repair and mutagenesis. Environ Mol Mutagen 58: 235-263.

[43] Vergas-Rondon N, Villegas V E, Ronon-Lagos M (2018). The role of chromosomal instability in cancer and therapeutic responses. Cancers 10: 4.

Chapter 5

Genome Modifications Involved in Developmental Programs of the Placental Trophoblast

Tatiana G. Zybina

Abstract

The placental trophoblast cells give an example of profound genome modifications that lead to whole-genome multiplication, aneuploidy, under-replication of some genes or their clusters as well as, by contrast, gene amplification. These events are included into program of differentiation of functionally different cell lineages. In some cases the trophoblast cell differentiation involves depolyploidization achieved by non-mitotic division. Aneuploidy may be also accounted for by the unusual mitoses characteristic of Invertebrates and plants; in mammalian it may result from hypomethylation of centromere chromosome regions. The giant (endopolyploid) trophoblast cells organization includes "loose nucleosomes" accounted for by the non-canonical histone variants, i.e. H2AX, H2AZ, and H3. 3 . In the human extravillous trophoblast cells that, like murine TGC, invade endometrium, there occured significant changes of methylation as compared to non-invasive trophoblast cell populations . Meantime, some genes show hypermethylation connected with start of trophoblast lineages specification. Thus, despite the limited possibilities of chromosome visualization trophoblast cells represent an interesting model to investigate the role of modification of gene copy number and their expression that is important for the normal or abnormal cell differentiation.

Keywords: trophoblast, genome, chromatin, chromosomes, histones, polyploidy, aneuploidy, gene amplification

1. Introduction

The placental trophoblast gives examples of genome multiplication included in the program of their lifespan during embryogenesis [1–8]. The degree of ploidy varies between different trophoblastic cell lineages and among different mammalian species indicating that increase of chromosomes or gene copies is required for functional activity of the cell or is dictated by the lifestiyle of a species [3, 5, 9]. Besides euploidy, the trophoblast cells show aneuploidy examples that varies between different mammalian species and may reflect both differences in the genome structure and, again, requirements of species-specific way of life. Here we would like to match the above-mentioned genome peculiarities with the cell nucleus organization of the trophoblast cells that may underlie their specific functions.

2. Poly- and aneuploidy, their origin and significance

Recently, a notion dominates that the multifold genome multiplication is achieved by modified cell cycles. Among them the shortest one leading to the highest levels of ploidy is characterized by alternating DNA synthesis (S) and Gap (G) phases in the absence of intervening mitoses, karyokinesis, and cytokinesis; a series of these shortened cycles allows cells to achieve high level of ploidy that may exceed 1000c [10–13].

The trophoblast cells that form a barrier between semiallogenic fetal (trophoblastic) and maternal (decidua) tissues probably require mechanism(s) to sustain maternal-fetal tolerance achieved by different mechanisms. For example, the trophoblast cells secrete a range of cytokines and chemokines thereby contributing to the process of immune regulation at the placental–maternal interface [14, 15]. On the other hand, as we stated previously, the TGC multifold genome multiplication also may protect their genome from mutagenic effect of the DNA of the phagocytosed maternal cells [16, 17]. Besides, some of the TGC functions of a barrier may be performed due to their giantism. TGC produce enormous keratin-positive sprouts that allow them to phagocytose accumulations of decidual cells and simultaneously to sustain the continuous TGC layer at the border with decidua [18]. Destruction of the cytokeratin 8 and 19 results in disruption of integrity of the murine giant trophoblast cell layer [19], which result in embryo death.

In distinct from the primary and secondary TGC, the low-ploid trophoblast cells in rat and mouse placenta show high proliferative activity and, being protected by a TGC barrier, accumulate a great bulk of cells that differentiate into a range of trophoblast cell subtypes, some of them form placental barrier supplying embryo by nutrition and oxygen; other subtypes are involved in glycogen storage, hormone production and deep intrauterine invasion [20–26].

The lifespan of the endoreduplicated TGC ends in depolyploidization via non-mitotic division of the giant nucleus or nuclear whole-genome fragmentation. In this case, division is achieved without complete chromosome condensation and their arrangement in metaphase plate, spindle formation and poleward chromosome movement. DNA content as well as nucleoli, heterochromatin and gonosomal chromatin bodies distributed into "subnuclei" according to their ploidy levels [3, 27–30]. By now, it is possible to consider it as variant of so called "polyploidy cycle" [31–33] that implies alternation of diploid and polyploid state in a cell lineage. It should be noted that such a phenomenon is fairly rare encountered in the cell lifespan and may be found in the "ancient" organisms like Protists [31–33] and some Invertebrates [34]. In the multicellular Invertebrates and Plants a wast majority of the differentiated cell types are endopolyploid [35, 36]. In Vertebrates, most cells are diploid, and the mammalian trophoblast cells, probably, represent an example of a recapitulation to some ancient forms of cell cycle and cell lifespan similar to protists and Invertebrates.

As to depolyploidization in TGC of rodent placenta, it should be emphasized that they do not belong to the complete polyploid cycle because they do not give rise to the cells capable of self-reproduction because they cease DNA replication shortly before the birth that probably prevents a massive proliferation of semiallogenic embryonic cells inside the maternal tissues.

As we stated in our previous paper, depolyploidization probably may result in aneuploidy in the trophoblast cells [5] because such a way of cell dividion, most probably does not ensure precise distribution of all chromosomes into dauther cells. Surprisingly, aneuploidy combined with polyploidy were recently reported as a factor of adaptation to the stressful conditions [37, 38]. Hepatocytes represent a cell type capable of high mitotic activity [39]. In the polyploid cells multipolar mitoses are encountered that may result in cells of lower ploidy, some of them were aneuploid [40]. Thus, in mice with knockdown of the genes *E2f7* and *E2f8* that regulate polyploidy in the liver, the amount of polyploid hepatocytes reduced fourfold; interestingly, nearly all hepatocytes became euploid [36]. Therefore, aneuploid cell resulted from polyploid ones. These mice were bred to tyrosinemia mice. As a result, although tyrosinemic mice were more susceptible to morbidities and death, they developed regenerating nodules similar to the control mice. Notably, the nodules in tyrosinemic livers were generated by aneuploidy; moreover, the mutation of *E2f7* and *E2f8* were inactivated [40]. The authors state that polyploid hepatocytes are necessary for the formation of aneuploid cells that can facilitate adaptation to chronic liver diseases.

In the placental trophoblast wide variability in different mammalian species show aneuploidy. In rodents, TGC of rat and mice undergone genome segregation via nuclear fragmentation, showed a great number of DNA values intermediate between ploidy classes [27]. It may be accounted for the deviation from the regular chromosome distribution rather than S-phase because at this developmental stage TGC do not proceed cell cycle and DNA replication that would be a reason of intermedial DNA content values. In contrast, in another rodent, field vole *Microtus rossiaemeridionalis*, TGC demonstrate the clear-cut classes of ploidy from 1c to 16 c [28]. Much more DNA content variability was found in the silver fox placenta [41, 42]. By contrast, silver fox placenta, especially its invasive trophoblast, shows a notable fluctuation of ploidy and a variability of patterns of polyploidization within the same cell lineage including aneuploidy and genome multiplication pathways (endoreduplication, classic endomitosis, depolyploidization).

It may reflect the necessity of different strategies that may be useful for maintaining the lengthy pregnancy [5, 42]. In the silver fox placenta, upon polyploidization, a considerable deviation from (2^n)c was found, with a tendency to $2^n \times 3$c, and there were a great variety of intermediate values suggesting a significant incidence of aneuploidy [41]. We suggested that it may serve a source of genome variability, in particular, hetero- and homozygosity that may be useful to select a more specific response to stress factors. Unlike small rodents such as mouse, rat and field vole, whose pregnancy do not exceed 30 days, in the fox placenta aneuploid trophoblast cells may have a protective effect during 6 months of intrauterine development [5].

3. Underreplication and amplification of some genes and clasters regulate the giant trophoblast cells differentiation

The cells undergone endoreduplication and formation of the classic polytene chromosomes are known to underreplicate a significant amounts of DNA [43, 44].

Recently it has been found underreplication (UR) of some chromosome regions and genes in the murine giant trophoblast cells. TGCs of the mouse placenta contain 47 regions, totaling 138 Megabases, where genomic copies are underrepresented [45]. UR domains originate from a subset of late-replicating heterochromatic regions containing gene deserts and genes involved in cell adhesion and neurogenesis. Interestingly, both size and degree of depletion of UR domains gradually progresses during early gestation. Thus, all UR domains at 9.5th day are also present at 8.0th day, and UR domains at 9.5 gestation day are also more numerous, larger and more depleted. However, unlike between 8.0th and 9.5th day, where the degree of depletion expanded, there were no significant change from 9.5th and 16.5th day.

Notably, 8-10th day of gestation in mice corresponds to the placenta formation whereas at 10-16th days well-developed placenta functioning takes place. The authors [45] note that the increase in UR domain size and degree of underrepresentation from 8.0th to 9.5th day is linked to the "robust" endocycles of early gestation [46].

Besides, it should be mentionned that, during the late stages of TGC lifespan, new underreplicated regions are also formed but they are more stochastic, less reproducible, and significantly smaller than those conserved between all stages [45]. Notably, underreplication of TGC coincide with period of the most significant stages of TGC invasion and anchoring to endometrium and is integrated in its developmental program.

The above-mentioned data also show that UR domains are formed from a specific class of late-replicating heterochromatic regions that contain mainly non-coding DNA, suggesting that UR domains are not simply a byproduct of late-replicating heterochromatin, but are a precisely regulated subset of DNA sequences. The authors come to conclusion that presence of UR domains in *Drosophila* endoreduplicated cells and in murine TGC is an example of convergent evolution. In this case UR contributes to accelerating the cell cycles that makes possible fast rate of development both in flies and in mice [45].

The underreplication in the endoreduplicated trophoblast cells not only may fasten the cell cycles but also be important for the TGC specific functions. Thus, Hannibal et al. [45] also note that UR domains are enriched for specific classes of genes involved in cell adhesion and neurogenesis. It is still difficult to find an explanation for the UR of specific genes and gene clusters. It can only be assumed that a certain number of gene copies is optimal in a given cell type. UR of separate genome regions at the background of its multiple duplication makes it possible to fine-tune the number of functioning copies necessary for performing specific functions.

In some cases, significance of UR was clearly demonstrated. Thus, downregulation of genes that regulate cell adhesion, junction and related cytoskeleton rearrangements is necessary for trophoblast EMT transition and invasion in both mice and humans [47, 48, 58]. Upregulation of genes in the SLIT/ROBO neuronal guidance system in the human placenta has been found to be bound with preeclampsia [49]. It is possible, placenta oxygenation requires precise specific function of SLIT/ROBO signaling achieved by UR of its genes.

Therefore, significance of endoreduplication is not only multifold genome duplication itself and enlargement of the cell that may be of significance for TGC barrier function but also a possibility of a fine regulation of a number of functional gene copies to provide cells capabilities to accomplish some functions necessary at the precise stages of development.

4. Amplification of some genes significant for the pregnancy also takes place in TGC

The mammalian polytene chromosomes may also undergo amplification of specific gene cluster. In the murine placenta TGC, five amplified regions were found using whole-genome sequencing and digital droplet PCR [50]. All the gene clusters are known to play key roles in mammalian placenta development and maintenance: the prolactins that regulate trophoblast cell lineage differentiations [51], serpins [52] and cathepsins [53] that promote trophoblast invasion, as well as (NK)/C-type lectin complex that play a crucial role in the feto-maternal cross-talk [54–56].

Therefore, amplification at selective genomic regions is another important mode of genome regulation in placental TGCs.

5. Unusual chromatin structure of TGC

Besides the non-classic polytene chromosomes in rodent placenta TGC [5, 26], some details of unusual chromosome structure have been revealed recently in the endoreduplicated TGC of mice. In the course of differentiation of TSC into TGC, expression of most genes encoding canonical histone were downregulated [1] By contrast, genes encoding non-canonical histones - H2AX, H2AZ and H3.3 did not show downregulation. The micrococcal nuclease digesion assay as well as nucleosome stability assay using a microfluidic devise showed that chromatin progressive loosening of chromatin in the course of TSC differentiated. Experiments combining H3.3 knockdown and overexpression showed that variant H3.3 resulted in formation of the loose nucleosomes in the murine TGC [1].

The presence of H2AZ and H3.3 in the genome potentially correlated with actively transcribed genes, indicating that H2AZ and H3.3 were necessary for creating relaxed and transcriptionally active chromatin structures [57–59]. Therefore, H2AX, H2AZ, and H3.3 histone variant may be responsible for the formation of a loose nucleosome structure that was unique to TGCs [1].

Interestingly, knockdown of H3.3 variant in the differentiationg TSCs significantly decreased the number of cells containing more than 4n DNA content compared to the control cells [1]. Therefore, switch to the non-canonic histone variants seems to be a prereqisite of the trophoblast cell endoreduplication. On the other hand, loose chromatin organization may, like underreplication, promote fastening the modified cell cycle that allow reach multifold (up to 512c and higher) genome multiplication and formation a giant trophoblast cell layer at the border with semiallogenic maternal tissue.

The unusual chromatin status is revealed, in particular, in the organization of the inactive X-chromosome of the murine TGC [60]. Thus, investigation of the precise temporal and lineage-specific X-inactivation status of several genes in postimplantation mouse embryos showed stable gene silencing in most lineages, with significant levels of escape from XCI mainly in one extra-embryonic cell type - TGCs. It has been found that the *Xist* RNA-coated X chromosome has a highly unusual chromatin content in TGCs, presenting both heterochromatic marks such as H3K27me3 and euchromatic marks such as histone H4 acetylation and H3K4 methylation. This unusual combination of silent and active features is likely to reflect, and might underlie, the partial activity of the X chromosome in TGCs. However, some key loci seem to require dosage compensation in TGC that probably points out to combination of the relaxed and silenced gene expression as a specific mode of gene activity regulation in a condition of chromatin unusual organization in TGC.

6. Hypomethylation of human and rodent placenta

Methylation status provides some new insight in the understanding of the trophoblast cell organization that underly their unique features. The human placental trophoblast shows general global hypomethylation [61]. It is possible that the loose-nucleosome structure of TGC in murine placenta and global hypomethylation in human placenta are similar phenomena. In human placenta, genome-wide hypomethylation, coupled with gene-specific hypermethylation of tumor-suppressor genes, is a common feature of human cancers [64]. Interestingly, the placenta parallels human cancers in both the overall decreased level of genomic DNA methylation and the specific hypermethylation of several tumor suppressor genes [61–64].

Such a parallel with carcinogenesis may be connected with the trophoblast invasive pathways. Inhibition of DNA methylation by 5-azacytidine treatment disrupts trophoblast invasive and migratory potential *in vitro* [65] and proper placental development *in vivo* [66]. Thus, threatment of BeWo cell with DNA methyltransferase inhibitor, 5′-aza-2′-deoxycytidine (AZA) resulted in conversion to non-migratory phenotype. AZA was found to increase mRNA level of E-cadherin and plakoglobin, components of cell junction structures - zonula adherens and desmosomes [65]. AZA treatment also resulted in decrease their gene promoter activity and protein levels. Increases in plakoglobin and E-cadherin promoter activity and inhibition of BeWo cell migration was also achieved with small interfering RNA-mediated depletion of both *DNMT-3a* and *DNMT-3b* [65].

Meantime, beside the trophoblast invasion, some DNA methylation (locus-specific and/or repeat-based) is important for differentiation of the functionally different trophoblast lineages forming the placental barrier and performing other placental specific functions. Thus, most homeobox genes were hypomethylated in the human placenta throughout gestation. Nevertheless, three homeobox genes, *TLX1*, *HOX10* and *DLX5* showed progressive methylation and decrease of their expression in the course of pregnancy – from first to third trimester. Using siRNA treatment the key role of *TLX1, HOXA10* and *DLX5* in trophoblast proliferation, differentiation and apoptosis was demonstrated [67]. It cannot be ruled out that progressive methylation of some homeobox genes promotes trophoblast differentiation into highly proliferative ones that, in turn, gives rise to villous cyto- and syncytiotrophoblast and the invasive trophoblast lineages. The data suggest an important role of several homeobox gene methylation in gene expression in the course of placenta formation.

In mice, *Dnmt3L* is expressed at high levels in the chorion, containing a multipotent trophoblast stem cell population. Disruption of *Dnmt3L* disturbs placental development including spongiotrophoblast and labyrinth malformation, leads to excess of TGC and defective attachment of the chorion to the ectoplacental cone. Excessive TGC development indicates that such a phenotype is bound to the hypomethylation. This is associated with an arrest of proliferation of the extraembryonic tissue [68]. It may be suggested that *Dnmt3L*-mediated *de novo* methylation is connected with initiation of differentiation of trophoblast into multiple lineages that imply maintenance of high level of mitotic proliferation regulated by Mash2 expression as well as syncytiotrophoblast formation demonstrated by *GCM1* expression [68]. It suggests that *Dnmt3L*-mediated *de novo* methylation is critical for proper placental development in mice [61].

Demonstration of placenta-specific hypomethylation of the *DNMT3L* gene supports a role in human placental development [61] probably bound to the start of differentiation of trophoblast into a range of lineages with different proliferative capacity.

7. Hypomethylation and chromosome decondensation may involve genome rearrangement

Thus, the trophoblast cell population are prone to hypomethylation that may be a prerequisite of some other cytogenetic effects. Thus, prolonged culture of normal chorionic villus cells involves chromatin decondensation and rearrangements that mimics the ICF syndrome, i.e. immunodeficiency, centromeric region instability, and facial anomalies [69]. Thus, untreated cultures from normal chorionic villus or amniotic fluid-derived samples displayed dramatic cell passage-dependent increases in aberrations in the juxtacentromeric heterochromatin of chromosomes 1 or 16

(1qh or 16qh). By passage 8 or 9, 82 ± 7% of the chorionic villi metaphases from all eight studied samples exhibited 1qh or 16qh decondensation and 25 ± 16% had rearrangements in these regions. At early and late passages, chorionic villi DNA was hypomethylated, and amniotic fluid DNA was hypermethylated both globally and at Sat2. *DNMT1*, *DNMT3A*, or *DNMT3B* RNA levels did not differ significantly between chorionic and amniotic cultures, or late and early passages. Sat2 hypomethylation may favor 1qh and 16qh anomalies because the chorionic villi cultures, with their Sat2 hypomethylation, displayed 1qh and 16qh decondensation Therefore, hypomethylation characteristic of the embryonic and trophoblast cells may affect the structure heterochromatin regions leading to chromosome rearrangement that may be a source of genome variability at the stage of differentiation of trophoblast cell population.

8. Conclusion

The data of the present paper indicate that the multifold genome multiplication is not the only peculiarity of the placental trophoblast cell lineages. There are a range of other peculiarities that, most probably, result from cell cycle modification or play a role connected with genome duplication. Thus, underreplication and gene amplification may result in fine orchestration of gene copies necessary at a definite trophoblast cell line in order to accomplish its function. "Loose nucleosomes", most probably, represent one of peculiarities of endoreduplication; it cannot be ruled out that lack of mitoses do not require chromosome to proceed all levels of condensation and packaging of chromosomes that makes possible fastening of replication cycles and high level of transcription of different factors regulating formation of the provisory organ – placenta. Unmethylation of chromosomes including centromere regions may contribute to chromosome rise of aberrations that may be a source of genetic variability and selection of the optimal genetic structure that may have an adaptive effect during pregnancy.

Acknowledgements

The author is grateful to the Institute of Cytology for the support of the work as well as Prof. N. Nikolsky and Dr. Grigori Stein for their helpful advices.

Conflict of interest

The author declares no conflict of interests.

Author details

Tatiana G. Zybina
Pavlov First Saint-Petersburg State Medical University, St.-Petersburg, Russian Federation

*Address all correspondence to: tatanazybina610@gmail.com

References

[1] Hayakawa K, Terada K, Takahashi T, Oana H, Washizu M, Tanaka S, Nucleosomes of polyploid trophoblast giant cells mostly consist of histone variants and form loose chromatin structure. Sci Rep. 2018; 8: 5811. doi: 10.1038/s41598-018-23832-2.

[2] Gamage TKJB, Schierding W, Hurley D, Peter Tsai P, Jackie L. Ludgate JL, Bhoothpurd C, Chamley LW, Weeks RJ, Macaulay EC, James JL, The role of DNA methylation in human trophoblast differentiation. Epigenetics. 2018; 13: 1154-1173. DOI:10.1080/15592294.2018.1549462

[3] Zybina T, Zybina E, Cell reproduction and genome multiplication in the proliferative and invasive trophoblast cell populations of mammalian placenta. Cell Biol Intern. 2005; 29: 1071-1083. DOI: 10.1016/j.cellbi.2005.10.015

[4] Zybina T, Zybina E, Genome variation in the trophoblast cell lifespan: Diploidy, polyteny, depolytenization, genome segregation. 2014; 4: 77-93. www.wignet.com

[5] Zybina TG, Zybina EV, Role of cell cycling polyploidy in placental trophoblast of different mammalian species. Reprod Dom Anim. 2020; 55:895-904. DOI: 10.1111/rda.13732

[6] Hu D, Cross JC. Development and function of trophoblast giant cells in the rodent placenta. Intern J Devel Biol. 2010; 54: 341-354. DOI: 10.1387/ijdb.082768dh

[7] Biron-Shental T, Fejgin M D, Sifakis S, Liberman M, Antsaklis A, Amiel A, Endoreduplication in cervical trophoblast cells from normal pregnancies. J Matern-Fet Neonat Med. 2012; 25:2625-2628. DOI:10.3109/14767058.2012.717999

[8] Velicky, P, Meinhardt, G, Plessl, K., Vondra, S., Weiss, T., Haslinger, P, Lendl T, Aumayr K, Marihofer M, Zhu X, Schutz B, Hannibal R, Lindau R., Weil B, Emerudh J, Neesen J, Edder G, Mikula M, Rohrl C, Urbam AE, Baker J, Knofler M, Pollheimer J. Genome amplification and cellular senescence are hallmarks of human placenta development. PLoS Genet. 2018; 12: e1007698. DOI: 10.1371/journal. Pgen.1007698

[9] Zybina, T.G., Zybina, E.V. Cell cycle modification in trophoblast cell populations in the course of placenta formation. In: DNA replication and related cellular processes. Rijeka, Croatia: InTech; 2011. pp. 227-258. www.intechopen.com

[10] Fox DT, Duronio RJ, Endoreplication and polyploidy: insights into development and disease. Development. 2013; 140: 3-12. DOI:10.1242/dev.080531

[11] Brodsky V J, Uryvaeva I V 1985. Genome multiplication in growth and development. Moskow: Univ. Press. 497 p.

[12] Edgar BA, Orr-Weaver TL, Endoreduplication cell cycle: more for less. Cell. 2001, 105: 297-306. DOI: 10.1016/s0092-8674(01)00334-8

[13] Stormo B, Fox DT, Polyteny: still a giant player in chromosome research. Chromosome Res. 2017; 25: 201-214. DOI: 10.1007%2Fs10577-017-9562-z

[14] Ramhorst R, Fraccaroli L, Aldo P, Alvero AB, Cardenas I, Leirós CP, Mor G. Modulation and recruitment of inducible regulatory T cells by first trimester trophoblast cells. Am J Reprod Immunol 2012; 67: 17-27. https://dx.doi.org/10.1111%2Fj.1600-0897.2011.01056.x

[15] Du, M. R. et al. Embryonic trophoblasts induce decidual regulatory

T cell differentiation and maternal–fetal tolerance through thymic stromal lymphopoietin instructing dendritic cells. J. Immunol. 192, 1502-1511 (2014). DOI: 10.4049/jimmunol.1203425.

[16] Soloveva V, Linzer DIH, Differentiation of placental trophoblast giant cells requires downregulation of p53 and Rb. Placenta 2004; 25: 29-36. https://doi.org/10.1016/S0143-4004(03)00215 -7

[17] Guershenson C M, Alexandrov Y N, Maluta RD, Mutagenous effect of DNA and viruses in Drosophila. Kiev: Naukova Dumka, 1975. p. 305

[18] Zybina TG, Stein GI, Zybina EV, Endopolyploid and proliferating trophoblast cells express different patterns of intracellular cytokeratin and glycogen localization in the rat placenta. Cell Biol Intern. 2011; 35:649-655. DOI:10.1042/CBI20100278

[19] Tamai Y, Ishikawa T, Bösl MR, Mori M, Nozaki M, Baribault H, Oshima RG, Taketo MM (2000) Cytokeratins 8 and 19 in the mouse placental development. J Cell Biol 151: 563-72.

[20] Adamson SL, Lu Y, Whiteley KJ, Holmyard D, Hemberger M, Pfarrer C, Cross, JC, Interactions between trophoblastcells and the maternal and fetal circulation in the mouse placenta. Developmental Biology. 2002; 250; 358-373. https://doi.org/10.1006/dbio.2002.0773

[21] Caluwaerts, S., Vercruysse, L., Luyten, C., & Pijnenborg, R, Endovascular trophoblast invasion and associated structural changes in uterine spiral arteries of the pregnant rats. Placenta. 2005; 26: 574-584. https://doi.org/10.1016/j.place nta.2004.09.007

[22] Eaton, M., Davies, A. H., Devine, J., Zhao, X., Simmons, D., Mariusdottir, E., Natale DRC, Matyas JR, Elizabeth A. Bering EA, Matthew L. Workentine ML, Hallgrimsson B, Cross, JC, Complex pattern of cell growth in the placenta in normal pregnancy and as adaptation to maternal diet restriction. PLoS One. 2020; 15(1), e0226735. DOI:10.1371/journal.pone.0226735

[23] Hemberger M, Nozaki T, Masutani M, Cross, JC, Differential expression of angiogenic and vasodilatory factors by invasive trophoblast giant cells depending on depth of invasion. Devel. Dynam.: Off Publications of American Association of Anatomy, 2003; 227: 185-191. DOI: 10.1002/dvdy.10291

[24] Rai, A, Cross JC, Development of the hemochorial maternal vascular spaces in the placenta through endothelial and vasculogenic mimicry. Devel. Biol. 2014; 387: 131-14. DOI: 10.1016/j.ydbio.2014.01.015

[25] Simmons D G, Rawn S, Davies, A, Hughes M, Cross JC, Spatial and temporal expression of the 23 murine Prolactin/Placental Lactogen-related genes is not associated with their position in the locus. BMC Genomics. 2008; 9: 352. DOI: 10.1186/1471-2164-9-352

[26] Soares MJ, Chakraborty D, Mohammad A, Rumi K, Konno T, Renaud SJ, Rat placentation: An experimental model for investigating the hemochorial maternal-fetal interface. Placenta. 2012; 33; 233-243. DOI: 10.1016/j.placenta.2011.11.026

[27] Zybina V, Kudryavtseva MV, Kudryavtsev BN, The distribution of chromosome material during giant nuclei division by fragmentation in the trophoblast of rodents. Morphological and cytophotometrical study. Tsitologiya, 1979 21: 12-20 (In Russian).

[28] Zybina TG, DNA in the nuclear fragments that appear in the course of fragmentation of the secondary giant

trophoblast cells in the field vole. Tsitologiya. 1990; 32: 806-810. (Russian).

[29] Zybina TG, Zybina E V, Polytene chromosomes in mammalian cells. Intern Rev Cytol. 1996; 165: 53-119

[30] Zybina EV, Zybina TG, Bogdanova MS, Stein GI. Whole-genome chromosome distribution during nuclear fragmentation of giant trophoblast cells of *Microtus rossiaemeridionalis* studied with the use of gonosomal chromatin arrangement. Cell Biol Intern. 2005; 29: 1066-1070. DOI: 10.1016/j.cellbi.2005.10.014

[31] Raikov IB, The protozoan nucleus. Morphology and evolution. Cell biology monographs 9. Berlin: Springer-Verlag. 1982.

[32] Berdieva M, Demin S, Goodkov A 2019, *Amoeba proteus* and ploidy cycles: From simple model to complex issues. Protistology. 2019; 12: 166-173. DOI: 10.21685 /1680-0826-2019-13-3-6

[33] Demin SY, Berdieva, MA, Goodkov AV, Cyclic polyploidy in obligate agamic amoebae. Cell Tiss. Biol. 2019; 13: 242-246. DOI: 10.1134/S1990 519X1 9030027

[34] Luchetta E, Ohlstein, B, Amitosis of polyploid cells regenerates functional stem cells in the *Drosophila* intestine. Cell Stem Cell. 2017; 20: 609-620, e6. DOI:10.1016/j.stem.2017.02.012

[35] Nagl W. Endopolyploidy and polyteny in differentiation and evolution. Toward an understanding of quantitative and qualitative variation of nuclear DNA in ontogeny and phylogeny. 1978. North Holland Publishing Company: Amsterdam – N.Y. – Oxford.

[36] Nagl W, Cdc-2 kinases, cyclins and the switch from proliferation to polyploidization. Protoplasma 1995; 188: 143-150.

[37] Duncan AW, Aneuploidy, polyploidy and ploidy reversal in the liver. Semin Cell Dev Biol. 2013; 24: 347-356. DOI: 10.1002/hep.26233

[38] Duncan AW, Taylor MH, Hickey RD, Hanlon Newell AE, Lenzi ML, Olson SB, Finegold MJ, Grompe M, The ploidy conveyor of mature hepatocytes as a source of genetic variation. Nature. 2010; 467: 707-710. DOI:10.1038/nature09414

[39] Kudryavtsev BN, Kudryavtseva MV, Sakuta GA, Stein GI. Human hepatocyte polyploidization kinetics in the course of life cycle. Virchows Arch B Cell Pathol Incl Mol Pathol. 1993; 64: 387-393 DOI: 10.1007/BF02915139

[40] Wilkinson PD, Alencastro F, Delgado ER, Leek MP, Weirich MP, Otero PA, Roy N, Brown WK, Oertel M, Duncan AW, Polyploid hepatocytes facilitate adaptation and regeneration to chronic liver injury. Am J Pathol. 2019; 189: 1241-1255. DOI: 10.1016/j.ajpath.2019.02.008

[41] Zybina EV, Kudryavtseva MV, Kudryavtsev BN, The distribution of chromosome material during giant nucleus division by fragmentation in the trophoblast of rodents. Morphological and cytophotometrical study. Tsitologiya 1979; 21:12-20 (Russian)

[42] Zybina TG, DNA in the nuclear fragments that appear in the course of fragmentation of the secondary giant trophoblast cells in the field vole. Tsitologiya 1990; 32: 806-810 (Russian)

[43] Nordman J, Orr-Weaver TL, Regulation of DNA replication during development. Development. 2012; 139:455-464. DOI:10.1242/dev.061838

[44] Schoenfelder KP, Fox DT, The expanding implications of polyploidy. J Cell Biol. 2015; 209:485-491. DOI: 10.1083/jcb.201502016

[45] Hannibal RL, Chuong EB, Rivera-Mulia JC, Gilbert D, Valouev A., Baker JC, Copy number variation is a fundamental aspect of the placental genome. PLoS Genet. 2014; 10: e1004290. doi:10.1371/journal.pgen.1004290

[46] Sakaue-Sawano A, Hoshida T, Yo M, Takahashi R, Ohtawa K, et al. Visualizing developmentally programmed endoreplication in mammals using ubiquitin oscillators. Development. 2013; 140: 4624-4632. DOI: 10.1242/dev.099226

[47] El-Hashash AHK, Kimber SJ. PTHrP induces changes in cell cytoskeleton and E-cadherin and regulates Eph/Ephrin kinases and RhoGTPases in murine secondary trophoblast cells. Dev Biol. 2006; 290: 13-31. DOI: 10.1016/j.ydbio.2005.10.010

[48] Kokkinos MI, Murthi P, Wafai R, Thompson EW, Newgreen DF (2010) Cadherins in the human placenta–epithelial–mesenchymal transition (EMT) and placental development. Placenta. 2010; 31: 747-755. DOI:10.1016/j.placenta.2010. 06.017

[49] Liao WX, Laurent L, Agent S, Hodges J, Chen DB (2012) Human Placental Expression of SLIT/ROBO Signaling Cues: Effects of Preeclampsia and Hypoxia. Biol Reprod. 2012; 86: 111. DOI: 10.1095/biolreprod.110.0881 38.

[50] Hannibal RL, Baker JC. Selective amplification of the genome surrounding key placental genes in trophoblast giant cells. Curr Biol. 2016; 26:230-236. DOI:10.1016/j.cub.2015.11.060

[51] Soares, M.J., Konno, T., and Alam, S.M. (2007). The prolactin family: effectors of pregnancy-dependent adaptations. Trends Endocrinol. Metab. 18, 114-121. DOI: 10.1016/j.tem.2007.02.005.

[52] Kaiserman, D., Knaggs, S., Scarff, K.L., Gillard, A., Mirza, G., Cadman, M., McKeone, R., Denny, P., Cooley, J., Benarafa, C., et al. Comparison of human chromosome 6p25 with mouse chromosome 13 reveals a greatly expanded ov-serpin gene repertoire in the mouse. Genomics 2002; 79, 349-362. 10.1006/geno.2002.6716

[53] Mason, R.W. (2008). Emerging functions of placental cathepsins. Placenta 29, 385-390. DOI: 10.1016/j.placenta.2008.02.006

[54] Chazara O, Xiong S, Moffett A, Maternal KIR and fetal HLA-C: a fine balance. J. Leukoc. Biol. 2011; 90; 703-716. DOI: 10.1189/jlb.0511227

[55] Drickamer K, Taylor ME, Recent insights into structures and functions of C-type lectins in the immune system. Curr. Opin. Struct. Biol. 2015, 34, 26-34. DOI: 10.1016/j.sbi.2015.06.003

[56] Hofer E, Sobanov Y, Brostjan C, Lehrach H, Duchler M, The centromeric part of the human natural killer (NK) receptor complex: lectin-like receptor genes expressed in NK, dendritic and endothelial cells. Immunol. Rev. 2001; 181: 5-19. DOI: 10.1034/j.1600-065x.2001.1810101.x

[57] Jin C, Felsenfeld G, Nucleosome stability mediated by histone variants H3.3 and H2A.Z. Genes Dev. 2007; 21: 1519-1529. DOI:10.1101/gad.1547707

[58] Jin, C, Zang C, Wei G, Cui K, Peng W, Zhao K, Felsenfeld G, H3.3/H2A.Z double variant-containing nucleosomes mark 'nucleosome-free regions' of active promoters and other regulatory regions. Nat Genet. 2009; 41; 941-945. DOI: 10.1038/ng.409

[59] Yukawa M, Akiyama T, Franke V, Mise N, Isagawa T, Suzuki Y, Vlahovicek K, Abe K, Aburatani H,

Aoki F, Genome-wide analysis of the chromatin composition of histone H2A and H3 variants in mouse embryonic stem cells. PLoS One. 2014; 9, e92689. DOI: 10.1371/journal.pone.0092689

[60] Corbel C, Diabangouaya P, Anne-Valerie Gendrel A.-V., Chow JC, Heard E, Unusual chromatin status and organization of the inactive X chromosome in murine trophoblast giant cells. Development. 2013; 140: 861-872. doi:10.1242/dev.087429

[61] Novakovic B, Wong NC, Sibson M, Ng H-K, Morley R, Ursula Manuelpillai U, Down T, Rakyan VK, Beck S, Hiendleder S, Roberts CT, Craig JM, Saffery R, DNA methylation-mediated down-regulation of DNA methyltransferase-1 (*DNMT1*) is coincident with, but not essential for, global hypomethylation in human placenta. J. Biol. Chem. 2010; 285: 9583-9493.

[62] Novakovic B, Saffery R, Placental pseudo-malignancy from a DNA methylation perspective: unanswered questions and future directions. Frontiers in Genetics/ Epigenomics and Epigenetics. 2013; 4: article 285.

[63] Chiu RWK, Chim SSC, Wong IHN, Wong CSC, Lee W-S, To KF, Tong JHM, Yuen RKC, Shum ASW, Chan JKC, Chan LYS, Yuen JWF, Tong YK, Weier JF, Ferlatte C, Leung TN, Lau TN, Lo KW, Lo YMD. Hypermethylation of RASSF1A in human and rhesus placentas. Am J Pathol. 2007; 170: 941-950 DOI:10.2353/ajpath.2007.060641

[64] Fuke C, Shimabukuro M, A Petronis A, Sugimoto J, Oda T, Miura K, Miyazaki T, Ogura C, Okazaki Y, Jinno Y, Age related changes in 5-methylcytosine content in human peripheral leukocytes and placentas: an HPLC-based study. Ann Hum Genet 2004; 68(Pt 3):196-204.doi: 10.1046/j.1529-8817.2004.00081.x.

[65] Rahnama F, Shafiei F, Gluckman PD, Mitchell MD, Lobie PE, Epigenetic regulation of human trophoblastic cell migration and invasion. Endocrinology 147: 5257-5283

[66] Serman, L., Vlahovicŕ, M., Sijan, M., Bulicŕ-Jakus, F., Serman, A., Sincicŕ, N., Matijevicŕ, R., Juricŕ-Lekicŕ, G., and Katusicŕ, A. The Impact of 5-Azacytidine on Placental Weight, Glycoprotein Pattern and Proliferating Cell Nuclear Antigen Expression in Rat Placenta. Placenta. 2007; 28: 803-811

[67] Novakovic B, Fournier T, Harris LK, James J, Claire T, Roberts CT,Hannah EJ, Yong HEJ, Kalionis B, Evain-Brion D, Ebeling PR,Wallace EM, Saffery R, Murthi P. Increased methylation and decreased expression of homeobox genes *TKX1, HOXA10* and *DLX5* in human placenta are associated with trophoblast differentiation Scientif. Rep. 2017; 7: 4523. DOI: 10.1038/s41598-017-04776-5

[68] Arima T, Hata K, Tanaka S, Kusumi M, En Li, Kato K, Shiota K, Sasaki H, Wake N, Loss of the maternal imprint in Dnmt3Lmat−/− mice leads to a differentiation defect in the extraembryonic tissue. Dev. Biol. 2005; 297: 361-373. DOI: 10.1016/j.ydbio.2006.05.003

[69] Tsien F, Fiala ES, Youn B, Long TI, Laird PW, Weissbecker K, Ehrlich M, Prolonged culture of normal chorionic villus cells yields ICF syndrome-like chromatin decondensation and rearrangements. Cytogenet Genome Res. 2002; 98: 13-21

Chapter 6

Next-Generation Sequencing Revealed that High Proportion of Human Embryos Resulted from Donor Eggs Are Segmental Chromosome Abnormal

Xiangli Niu, Yanping Lao, Yan Sun and Weihua Wang

Abstract

High proportion of human embryos produced by in vitro fertilization (IVF) are aneuploidy or have segmental chromosomal errors. Not only a whole chromosome aneuploidy, but also small errors in a chromosome, such as microdeletion can be detected by current next-generation sequencing (NGS) for preimplantation genetic testing (PGT). The prevalence of aneuploidy in donor egg IVF was significantly different between fertility clinics. In the present study, we examined whether different embryo biopsy procedures affect embryonic aneuploidies in donor egg IVF. We did not find significant differences in the samples with abnormal chromosomes between two biopsy methods. When we further analyzed the samples with abnormal chromosomes, we found that 64.0–80.7% of the abnormalities were whole chromosome aneuploidies while 19.3–36.0% were segmental chromosome abnormalities. High embryo implantation rates were obtained after transferring screened euploid blastocysts. These results indicate that blastocyst biopsy procedures may not significantly affect embryo's chromosomal status, but PGT by high-resolution NGS revealed that high proportions of human embryos derived from donor eggs are not only aneuploidy, but also segmental chromosome abnormal, and screening of small chromosomal errors by NGS is beneficial to patients who use donated eggs for infertility treatment.

Keywords: aneuploidy, blastocyst, donor eggs, human, in vitro fertilization, preimplantation genetic testing, segmental chromosome abnormalities

1. Introduction

Embryonic aneuploidy in human in vitro fertilization (IVF) is very common and is one of the factors reducing embryo implantation and causing birth defects. Although aneuploidy is mainly observed in the embryos from patients with advanced maternal ages [1–3], it is also very common in the embryos from young patients and oocyte donors [4–9]. The frequencies of aneuploid embryos produced by IVF have been widely studied [4–12] by examination of all chromosomes with microarray and next generation of sequencing (NGS) through preimplantation genetic testing for

IntechOpen

aneuploidies (PGT-A) [13–17]. With PGT-A by NGS, not only a whole chromosome aneuploidy can be detected, but also segmental chromosome abnormalities (deletion and duplication) can be detected [18–21]. Segmental chromosome abnormalities typically represent regional losses or gains in one or more chromosomes. The size of a segmental abnormalities detectable by current NGS platforms is as small as 1 Mb, however, for PGT-A, usually 10 Mb and above are detected and reported.

Some segmental chromosome abnormalities may cause miscarriage and birth defect, while others may result in developmental delay and/or intellectual disability if the transfer of such embryos produce live birth. It has been found that the prevalence of embryonic aneuploidy in donor egg IVF was significantly different between fertility clinics indicating that clinical and laboratory procedures may be related to the occurrence of embryonic aneuploidies [12]. Embryo biopsy is a complicated and invasive laboratory procedure that involves several embryo manipulations during culture, so it may affect embryo's quality including aneuploidies. Therefore, in the present study, to examine whether embryo biopsy procedures affect embryonic aneuploidies in donor egg IVF, blastocysts were biopsied by two different biopsy methods and then the samples were examined by NGS. Collected data were analyzed in terms of the rates of embryos with whole chromosome aneuploidies and segmental chromosome abnormalities. Clinical outcomes, such as pregnant rate, live birth rate and embryo implantation rate were also analyzed.

2. Materials and methods

2.1 Ethical statement

The patients signed the consents for all laboratory and clinical procedures including embryo biopsy for PGT-A. The data was collected from medical records at the clinic and laboratory, and the study with PGT-A was approved by New England Institutional Review Board (NEIRB 14–504).

2.2 Donor stimulation

Donors for IVF treatment underwent controlled ovarian stimulation with a combination of daily injection of 75–300 IU recombinant follicle-stimulating hormone (Gonal-F, EMD Serono, MA, USA) and 75–300 IU of a combination of follicle stimulating hormone and luteinizing hormone (Menopur, Ferring Pharmaceuticals, NJ, USA). On day 5–7, 0.25 mg gonadotropin releasing hormone antagonist (Cetrotide, EMD Serono) was given daily until triggering for oocyte maturation by gonadotropin-releasing hormone agonist (Lupron) or human chorionic gonadotropin (hCG). Oocytes were retrieved at 35–36 hours after the trigger and then cultured in Global™Total medium (Origio Inc., CT, USA) at 37°C in an atmosphere of 5.5% CO_2, 6% O_2, and balanced N_2 under humidified or dry conditions.

2.3 Oocyte insemination and embryo culture

Oocytes were inseminated by intracytoplasmic sperm injection (ICSI) after cumulus cells were removed by using hyaluronidase (Fujifilm-Irvine Scientific) at 3–4 hours after oocyte retrieval and metaphase II oocytes were injected 4–5 hours after retrieval. After insemination, oocytes were cultured in Global™Total medium at 37°C in an atmosphere of 5.5% CO_2, 6% O_2, and balanced N_2 under humidified or dry conditions.

Fertilization was assessed 16–18 hours after insemination, and normal fertilization was characterized by two distinct pronuclei and two polar bodies. Fertilized oocytes were further cultured in the Global™Total medium and embryo quality was evaluated by an inverted microscope on day 3, 5, or 6.

2.4 Blastocyst biopsy

Two biopsy methods were used in the present study. The first is a traditional two-step method in which a small hole in zona pellucida was opened by laser pulses on cleavage embryos at Day 3. As shown in **Figure 1**, when embryo developed to blastocyst at day 5 or later and some cells from blastocysts hatched from the hole, 5–10 cells were aspirated to a biopsy pipette and then cells were separated from blastocyst proper by mechanical pulling and laser pulses.

The second is a modified and simplified one-step method with less embryo manipulation and less laser application. Hole opening in the zona pellucida was not performed on day 3 embryos. Blastocysts were directly processed for biopsy. The details for this method are as the follows: As shown in **Figure 2**, blastocyst for biopsy was held to a proper position (**Figure 2A**) in which inner cell mass (ICM) was on the 6–9 O'clock position, a small hole in the zona pellucida was opened (**Figure 2B**) on the 3 O'clock position by one laser pulse with the ZILOS-tk™ laser system (Hamilton Thorn Bioscience Inc., MA, USA). A 20 μm polished biopsy pipette (Sunlight Medical, Jacksonville, FL, USA) was inserted to inside zona pellucida through the hole and a few trophectoderm cells on the 12–2 O'clock position were aspired into biopsy pipette (**Figure 2B**). After biopsy pipette was pulled out of the zona (**Figure 2C**), biopsy pipette together with blastocyst was moved to the top the holding pipette (**Figure 2D**), and the biopsy pipette was pull down against the holding pipette so that the cells inside the biopsy pipette were completely separated from blastocyst proper (**Figure 2E**).

After biopsy, biopsied cells were washed individually, transferred to PCR tubes, and stored at −20°C freezer until processing for PGT-A by commercial genetic testing company. Blastocysts were individually cryopreserved by vitrification for later frozen embryo transfer (FET). Blastocysts were classified as abnormal if they

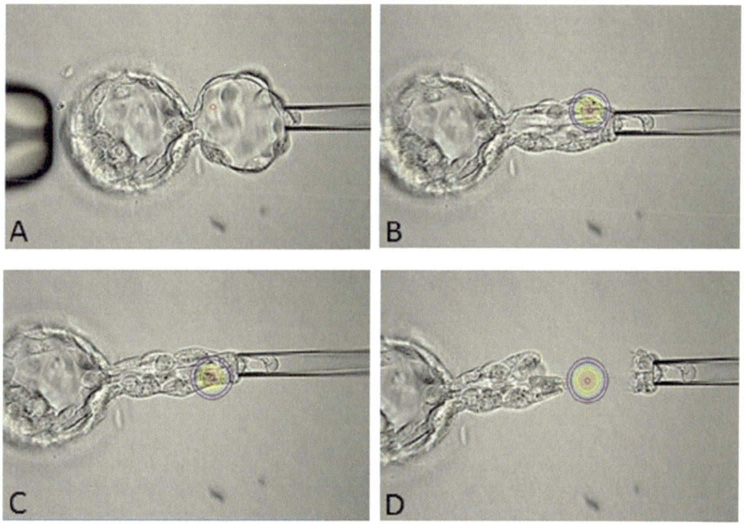

Figure 1.
Procedures for two-step blastocyst biopsy. A blastocyst with some trophectoderm cells being hatched from the hole in the zona pellucida opened on day 3 and the blastocyst is held to a proper position for biopsy (A). After a few trophectoderm cells are aspirated into biopsy pipette, one laser pulse is applied on upside of the cells (B) and another lase pulse is applied to the bottom side of cells (C) during mechanical pulling. Extra laser pulses may be necessary during pulling until the cells are completely isolated from blastocyst proper (D).

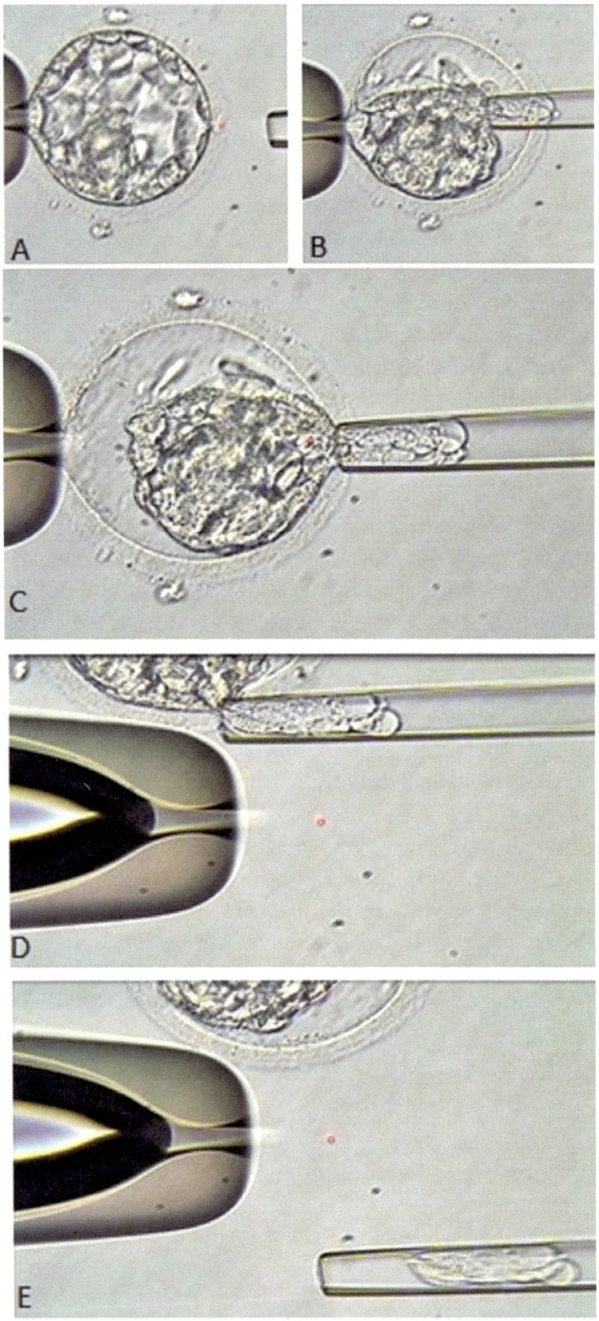

Figure 2.
Procedures for one-step blastocyst biopsy. A blastocyst is held to a proper position for biopsy and a small hole is opened in the zona by a laser pulse at 3 O'clock position (A). A biopsy pipette is inserted into the zona through the hole and a few trophectoderm cells are aspirated into biopsy pipette after blastocyst is collapsed or during collapsing from 12 to 2 O'clock position (B). Biopsy pipette is pull out of the zona (C) and the biopsy pipette together with blastocyst is moved to the top front of holding pipette (D). Biopsy pipette is pull down against the holding pipette to cut the cell connection between blastocyst proper and aspirated cells at the tip of biopsy pipette (E).

had any chromosomal error(s). The abnormal samples were further divided into aneuploidy if they had gain and/or missing of a chromosome(s) and segmental abnormal if they had only deletion and/or duplication in a chromosome(s).

2.5 Blastocyst vitrification, warming and transfer

Blastocysts were vitrified using a vitrification device (Cryotop, Vitristraw or Mini straw) and kit (Fujifilm-Irvine Scientific, Irvine, CA, USA). Both equilibration solution and vitrification solution were warmed in original vials at 37°C for at least 30 min before use. Briefly, collapsed blastocysts were equilibrated in 100 μl drop (without oil cover) of equilibration solution for 2 min, and then 45 seconds in 100 μl drop (without oil cover) of vitrification solution (both steps were performed on a 37°C warming stage) before loading to vitrification device. All blastocysts were vitrified individually and then stored in liquid nitrogen until warming for FET.

For warming, blastocysts were exposed to a thawing solution (Fujifilm-Irvine warming kit) at 37°C for 1 min, transferred to a dilution solution for 3 min and finally to a washing solution for 10 min with a solution change after 5 min at room temperature. After completion of the warming process, zona pellucida in the blastocysts were further cut by laser pulses to open 1/4–1/5 (2D image size) of zona pellucida and then cultured in Global™Total medium for 2–4 h before transfer.

For preparation of the transfer, patients received estradiol (Estrace, Warner Chilcott, NJ, USA) orally or vaginally, and estradiol patch (Estradiol Transdermal System, Noven Pharmaceuticals, NJ, USA) every three days, as well as progesterone that was administered on 15th day of estradiol treatment. Blastocysts were transferred on the sixth or seventh day of progesterone administered, and progesterone was continued daily until the first serum β-hCG test two weeks after transfer. Ongoing pregnancy was supported by continued estradiol and progesterone until 11 weeks of pregnancy. Pregnancy was initially confirmed 14 days after embryo transfer by a serum β-hCG assay. Four weeks after embryo transfer, when a gestational sac and a heartbeat appeared, the patient was diagnosed as having a clinical pregnancy. Live birth rates were calculated based on the number of live birth and number of transfers.

2.6 Statistical analysis

Interval data was analyzed by one-way analysis of ANOVA. The differences between groups were compared with chi square test. If the P value was less 0.05, the difference was considered to be statistically significant.

3. Results

To examine whether day 3 zona hole opening by laser pulse affected embryo development, blastocyst development between embryos with or without this procedure were compared. As shown in **Table 1**, similar blastocyst development rates (64.7 vs. 64.3%) were observed between two groups. Other parameters, such as egg donor's ages (26.5 ± 3.0 vs. 25.6 ± 2.6), and fertilization rates (86.4 vs. 88.8%) were also similar between two groups.

As shown in **Table 2**, after biopsy, the proportions of samples without tested results due to low quantity of DNA or no DNA in the samples were similar between two biopsy methods (3.6 vs. 4.4%), resulting in 96.4% of the samples biopsied with one-step method and 95.6% of the samples biopsied with two-step method were successfully amplified. It was found that euploid blastocyst rates were similar between two groups (63.4 vs. 64.0%).

Chromosome abnormalities include whole chromosome aneuploidies (extra and/or missing chromosomes), and segmental chromosome abnormalities, such as chromosome deletion and duplication. As shown in **Table 2**, no differences were

	Zona hole opening at Day 3 embryo		
	-	+	P Value
# of cases	61	45	NA
Donor age (Mean ± SD)	26.5 ± 3.0	25.6 ± 2.6	0.45
No. of eggs inseminated	1050	726	NA
No. of eggs fertilized (%)	907 (86.4)	645 (88.8)	0.12
No. of blastocysts (%)	587 (64.7)	415 (64.3)	0.88

NA: Not applicable.

Table 1.
Development of human embryos with or without laser zona hole opening that was performed on cleavage stage embryos at day 3.

	One-step method	Two-step method	P value
# of blastocysts biopsied	527	407	NA
# of samples without test results (%)	19 (3.6)	18 (4.4)	0.53
# of samples with test results (%)	508 (96.4)	389 (95.6)	0.53
# of euploid blastocysts (%)	322 (63.4)	249 (64.0)	0.85
# of embryos with abnormal chromosomes	186 (36.6)	140 (36.0)	0.85
# of aneuploid blastocysts (%)	119 (64.0)	113 (80.7)	0.06
# of segmental abnormalities (%)	67 (36.0)	27 (19.3)	0.06
# of samples with ≥2 abnormal chromosomes	45 (37.8)	50 (44.2)	0.32

NA: Not applicable.

Table 2.
Comparison of chromosomal abnormalities in the blastocysts after biopsy by one-step and two-step methods.

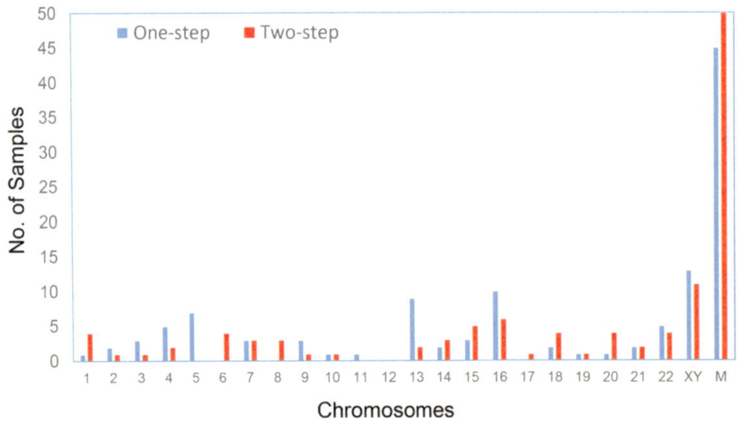

Figure 3.
Distribution of chromosomes in the aneuploid blastocysts biopsied with one-step and two-step method. Data represent the number of samples with a single abnormal chromosome and multiple (M: ≥2) abnormal chromosomes.

observed in the samples with whole chromosome aneuploid rates or segmental abnormalities between two biopsy methods. Samples with multiple chromosomal abnormalities were also similar between two biopsy methods.

	One-step method	Two-step method	P value
# of transfers	47	41	NA
Mean age of recipients	42.0 ± 6.8	42.9 ± 6.7	0.24
# of chemical pregnancy (%)*	37 (78.7)	26 (63.4)	0.11
# of clinical pregnancy (%)**	35 (74.5)	25 (61.0)	0.18
# of live birth (%)	33 (70.2)	24 (58.5)	0.25
# of blastocysts transferred	59	63	NA
# of blastocyst implanted	37(62.7)	31 (49.2)	0.11

*Positive beta hCG.
**Fetus with heartbeat.
NA: Not applicable.*

Table 3.
Comparison of clinical outcomes after transfer of euploid blastocysts biopsied by one-step and two-step methods from donor egg cycles.

As shown in **Figure 3**, aneuploidies occurred in all chromosomes except chromosome 12 and the differences for each chromosome were not statistically significant between two methods.

As shown in **Table 3**, transfer of euploid blastocysts biopsied by one-step method had higher chemical pregnancy (78.7 vs. 63.4%), clinical pregnancy (74.5 vs. 61.0%), live birth (70.2 vs. 58.5%) and embryo implantation rates (62.7 vs. 49.2%) as compared with transfer of euploid blastocysts biopsied by two-step method. Although these rates were not statistically significant between two groups, improved clinical outcomes were observed when one-step biopsy method was used.

4. Discussion

Recently Munne *et al.* found that euploidy rate in human embryos produced by donor egg IVF differed significantly between infertility clinics [12], but they did not analyze the cause(s) related to these differences. Because they collected data from multiple IVF clinics and each clinic used different biopsy methods, which may make it difficult to analyze these factors. In the present study, to minimize the effects of maternal age-related aneuploidy formation in the embryos [6, 13–15, 17–20], we also used donor egg IVF cycles to examine whether biopsy methods affect embryo aneuploidies. Our data indicate that biopsy methods do not affect embryonic aneuploidies, however, simplified biopsy method may improve embryo implantation that may be benefited from reduced embryo manipulations and limited laser applications. We also found that high proportions of human embryos from donor egg IVF are not only whole chromosome aneuploidy, but also have segmental chromosome abnormalities.

Although most of embryonic aneuploidies have already occurred before oocyte and sperm are collected for IVF due to meiotic error(s) during oocyte and sperm development [1–3], some of aneuploidies may be caused by mitotic errors, or suboptimal in vitro conditions and/or in vitro manipulations [1]. Rigorous temperature control during oocyte manipulations can maintain meiotic spindle integrity that may prevent meiosis error during final oocyte maturation after egg retrieval [1]. While embryo biopsy for PGT-A is still an invasive laboratory procedure, thus different methods may affect the embryo quality including chromosome integrity. Since laser was used to zona hole opening and blastocyst biopsy in human IVF, it has made the biopsy procedure to be easy [21]. However, excess use of laser pulses may be harmful to embryos and eventually would affect embryo development. For example, laser pulse(s) are applied on both cleavage embryos and blastocysts for the traditional two-step biopsy. The blastomeres next to the laser pulses may have different degrees of heat injuring, some injuries can be seen immediately or after further culture, while minor injuries may not be able to see under microscope during culture. Because the biopsied cells are mostly originated from these cells next to the position with laser pulses, chromosomes in some of these cells may be affected, which would eventually increase the rates of chromosomal abnormalities. However, based on our results observed in the present study, these manipulations of embryos do not affect chromosome integrity in the biopsied cells thus the aneuploidies after two-step biopsy were not increased as compared with one-step biopsy method. These results indicate that the chromosome abnormalities during embryo development, if occurs, are not from biopsy procedures.

However, biopsy procedures did affect embryo implantation. Previous studies with non-donor egg IVF found that blastocyst rates and live birth rates were reduced when day 3 zona opening was performed for two-step biopsy [22, 23]. Although we did not find the reduced blastocyst development after day 3 embryo manipulation in the present study, both embryo implantation and live birth rates were reduced when two-step biopsy procedure was used. These results indicate that laser pulses for zona hole opening at day 3 embryos may have detrimental effects on subsequent embryo development and embryo implantation ([22, 23], current study). We did not observe the differences in the blastocyst development after day 3 embryo manipulation in the present study as compared with no day 3 embryo manipulation, which may be due to good quality of oocytes from donors as compared oocytes from patients, thus some blastomeres might be affected, but overall blastocyst development rate was not reduced. Zona hole opening on day 3 embryos by laser pulses may affect embryo development especially if the perivitelline space is small or laser power is too large, thus the detrimental effects were caused by over-heating from laser pulses.

Although the statistical differences of embryo implantation rates between two biopsy methods were not significant due to small cycle numbers in the study by Zhao et al. [22] and in our current study, the differences were significant in the study by Rubino et al. in which more IVF cycles were examined [23]. When we reviewed the clinical outcomes by one-step and two-step biopsy methods in these studies, we found that the overall live birth rates could be increased by approximately 10% (9.26–12.7%) if one-step biopsy method was used, irrespective of small number of cycles or large number of cycles ([22, 23] and the current study) were analyzed.

Another reason for reduced embryo implantation after two-step biopsy may be resulted from blastocyst biopsy procedures. The traditional two-step blastocyst biopsy is performed by mechanical pulling and laser pulses. Heating from laser pulses would also cause injuries to the cells exposed to laser pulses, which would negatively affect embryo quality. Our one-step biopsy procedure is similar as that reported previously [22–24] but some modifications has been made. Cells were aspirated inside zona pellucida that is same as that used by Rubino et al. [23]. However, the cells aspirated into biopsy pipette were separated from blastocyst proper by mechanical blunt dissection, not by mechanical pulling and laser pulses, which is same as that reported by Zhao et al. [24].

The summarized benefits of our method are as the follows: First, one-step method does not need to have embryos to be exposed to laser pulse at day 3, which has been found to be detrimental to blastocyst development [22, 23]. Second, trophectoderm cells are aspirated inside zona pellucida, so that the fertilization of oocytes for PGT-A can be performed by either ICSI or regular IVF, and the contamination by cumulus cells or sperm can be minimized and avoided. Third, the separation of testing cells from blastocyst proper is made by mechanical blunt dissection, not by mechanical suction/pulling and multiple laser pulses, thus the further injuries by laser pulses on isolated cells and blastocyst proper can be avoided. And the last, ICM may hatch from the hole in some blastocysts if zona opening is done at day 3 embryos, thus biopsy need to be done on a different position [23], which would further affect embryo's implantation competence.

A previous study reported monozygotic twins when two-step biopsy was used [23]. In the present study, we did not observe any monozygotic twins after transfer of blastocysts biopsied by either one-step or two-step method, and this may be attributed to the zona cutting (1/4–1/5) in all frozen blastocysts after warming. We performed blastocyst vitrification after blastocysts were completely collapsed, and the blastocysts were still at collapsed status after warming. The perivitelline space was still very wide after warming, thus laser cutting of a large portion of zona pellucida did not have any injury to the blastocysts. This procedure may avoid monozygotic twins after blastocyst biopsy.

Although biopsy procedures did not affect aneuploid formation in donor egg IVF, the proportions of embryos from donor egg IVF with chromosomal abnormalities are very high [8, 9, 12, 14, 15]. In the present study, we found that these chromosome abnormalities include the whole chromosome aneuploidies and segmental chromosome abnormalities. It has been estimated that ~32% of segmental abnormalities are originated from meiosis [25]. However, most segmental abnormalities originate from mitosis and are present in a mosaic pattern [25, 26]. It has been found that segmental abnormalities can occur in any chromosome, and the frequency of deletions and duplications is roughly equal [25].

It has been reported that approximately ~6–15% of blastocysts from human IVF have segmental abnormalities when evaluated by current PGT-A methods with different analysis platforms [25, 27, 28]. The incidence of blastocysts with only segmental abnormalities is about 2.4%–7.5% of all samples examined [25, 27–29]. However, in the present study, the segmental chromosome abnormalities accounts for approximately 20–40% of the abnormalities, or around 6–12% of all samples examined, which were higher than previous reports [25, 27, 28]. These differences may be attributed to different PGT analysis platforms because the resolutions and accuracies are different between platforms. The high-resolution PGS platforms, such as NGS, can detect more small chromosome errors than previous microarray and low-resolution platforms. This may also be explanation that PGT by NGS improves pregnancy outcomes compared with array comparative genomic hybridization in single thawed euploid embryo transfer cycles [16].

It has been reported that the incidence of segmental abnormalities in human embryos do not correlate with patient age [25, 27, 28]. This may be the reason that high rates were observed in the embryos derived from young and healthy egg donors. Transfer of these embryos would result in failed implantation, miscarriage, or possibly liveborn congenital syndromes if carried to term [29, 30]. Some syndromes and conditions may be related to development delay and intellectual disabilities such as 1q21.1 deletion syndrome, 16p11.2 deletion syndrome and 1p36 deletion syndrome [31], thus screening of these syndromes that have small segmental chromosome abnormalities may be necessary in human IVF.

5. Conclusions

In conclusion, the outcomes from two previous studies [22, 23] and the current study indicate that one-step blastocyst biopsy can improve blastocyst implantation rate and live birth rate by ~10% in non-donor IVF patients [22, 23] and donor egg IVF patients (current study), suggesting that one-step biopsy method is superior to two-step method. Although blastocyst biopsy procedures may not affect the incidence of aneuploidies and/or segmental chromosome abnormalities, they affect embryo's implantation competences. Current PGT by high-resolution NGS reveals that high proportions of human embryos derived from donor eggs are not only whole chromosome aneuploidies, buy also segmental abnormal. Therefore, screening of these chromosome abnormalities may reduce embryo implantation failure, early miscarriage, birth defect, developmental delay and/or intellectual disability.

Acknowledgements

Supported by a grant from Guangxi Natural Science Foundation under Grant No. 2019GXNSFAA185016.

Conflict of interest

The authors declare no conflict of interest.

Author details

Xiangli Niu[1], Yanping Lao[1], Yan Sun[1] and Weihua Wang[2]*

1 Research Center for Reproductive Medicine, Reproductive Hospital of Guangxi Zhuang Autonomous Region, Nanning, Guangxi, China

2 Houston Fertility Laboratory, Houston, TX, USA

*Address all correspondence to: wangweihua11@yahoo.com

References

[1] Wang WH, Sun QY. Meiotic spindle, spindle checkpoint and embryonic aneuploidy. Front Biosci 2006; 11: 620-636.

[2] Jones KT. Meiosis in oocytes: predisposition to aneuploidy and its increased incidence with age. Hum Reprod Update 2008; 14:143-158.

[3] Hassold T, Hunt P. To err (meiotically) is human: the genesis of human aneuploidy. Nat Rev Genet 2001; 2: 280-291.

[4] Liang L, Wang CT, Sun X, Liu L, Li M, Witz C, Williams D, Griffith J, Skorupski J, Haddad G, Gill J, Wang WH. Identification of chromosomal errors in human preimplantation embryos with oligonucleotide DNA microarray. PLoS One 2013; 8: e61838.

[5] Baart EB, Martini E, van den Berg I, Macklon NS, Galjaard RJH, Fauser BCJM, Van Opstal D. Preimplantation genetic screening reveals a high incidence of aneuploidy and mosaicism in embryos from young women undergoing IVF. Hum Reprod 2006; 21: 223-233.

[6] Yang Z, Liu J, Collins GS, Salem SA, Liu X, Lyle SS, Peck AC, Sills ES, Salem RD. Selection of single blastocysts for fresh transfer via standard morphology assessment alone and with array CGH for good prognosis IVF patients, results from a randomized pilot study. Mol Cytogenet 2012; 5: 24.

[7] Munné S, Sandalinas M, Magli C, Gianaroli L, Cohen J, Warburton D. Increased rate of aneuploid embryos in young women with previous aneuploid conceptions. Prenatal Diagn 2004; 24: 638-643.

[8] Sills ES, Li X, Frederick JL, Khoury CD, Potter DA. Determining parental origin of embryo aneuploidy: analysis of genetic error observed in 305 embryos derived from anonymous donor oocyte IVF cycles. Mol Cytogenet 2014; 7: 68.

[9] Haddad G, Deng M, Wang CT, Witz C, Williams D, Griffith J, Skorupski J, Gill J, Wang W-H. Assessment of aneuploidy formation in human blastocysts resulting from donated eggs and the necessity of the embryos for aneuploidy screening. J Assist Reprod Genet 2015; 32: 999-1006.

[10] Franasiak JM, Forman EJ, Hong KH, Werner MD, Upham KM, Treff NR,Scott RT. The nature of aneuploidy with increasing age of the female partner: a review of 15,169 consecutive trophectoderm biopsies evaluated with comprehensive chromosomal screening. Fertil Steril 2014; 101: 656-663.

[11] Milán M, Cobo AC, Rodrigo L, Mateu E, Mercader A, Buendía P, Peinado V, Delgado A, Mir P, Simón C, Remohí J, Pellicer A, Rubio C. Redefining advanced maternal age as an indication for preimplantation genetic screening. Reprod Biomed Online 2010; 21: 649-657.

[12] Munné S, Alikani M, Ribustello L, Colls P, Martínez-Ortiz PA, McCulloh DH. Euploidy rates in donor egg cycles significantly differ between fertility centers. Hum Reprod 2017; 32: 743-749.

[13] Staessen C, Platteau P, Van Assche E, Michiels A, Tournaye H, Camus M, Devroey P, Liebaers I, Van Steirteghem A. Comparison of blastocyst transfer with or without preimplantation genetic diagnosis for aneuploidy screening in couples with advanced maternal age: a prospective randomized controlled trial. Hum Reprod 2004; 19: 2849-2858.

[14] Coates A, Bankowski BJ, Kung A, Griffin DK, Munne S. Differences in pregnancy outcomes in donor egg frozen embryo transfer (FET) cycles following preimplantation genetic screening (PGS): a single center retrospective study. J Assist Reprod Genet 2017; 34: 71-78.

[15] Lee E, Illingworth P, Wilton L, Chambers GM. The clinical effectiveness of preimplantation genetic diagnosis for aneuploidy in all 24 chromosomes (PGD-A): systematic review. Hum Reprod 2015; 30: 473-483.

[16] Friedenthal J, Maxwell SM, Munné S, Kramer Y, McCulloh DH, McCaffrey C, Grifo JA. Next generation sequencing for preimplantation genetic screening improves pregnancy outcomes compared with array comparative genomic hybridization in single thawed euploid embryo transfer cycles. Fertil Steril 2018; 109: 627-632.

[17] Liu J, Wang W, Sun X, Liu L, Jin H, Li M, Witz C, Williams D, Griffith J, Skorupski J, Haddad G, Gill J. DNA microarray reveals that high proportions of human blastocysts from women of advanced maternal age are aneuploid and mosaic. Biol Reprod 2012; 87: 1-9.

[18] Lee H-L, McCulloh DH, Hodes-Wertz B, Adler A, McCaffrey C, Grifo JA. In vitro fertilization with preimplantation genetic screening improves implantation and live birth in women age 40 through 43. J Assist Reprod Genet 2015; 32: 435-444.

[19] Munné S, Chen S, Fischer J, Colls P, Zheng X, Stevens J, Escudero T, Oter M, Schoolcraft B, Simpson JL, Cohen J. Preimplantation genetic diagnosis reduces pregnancy loss in women aged 35 years and older with a history of recurrent miscarriages. Fertil Steril 2005; 84: 331-335.

[20] Pagidas K, Ying Y, Keefe D. Predictive value of preimplantation genetic diagnosis for aneuploidy screening in repeated IVF-ET cycles among women with recurrent implantation failure. J Assist Reprod Genet 2008, 25,103-106.

[21] Schoolcraft WB, Fragouli E, Stevens J, Munne S, Katz-Jaffe MG, Wells D. Clinical application of comprehensive chromosomal screening at the blastocyst stage. Fertil Steril 2010; 94: 1700-1706.

[22] Zhao H, Tao W, Li M, Liu H, Wu, K, Ma S. Comparison of two protocols of blastocyst biopsy submitted to preimplantation genetic testing for aneuploidies: a randomized controlled trial. Arch Gynecol Obstet 2019; 299: 1487-1493.

[23] Rubino P, Tapia L, Alonso RFA, Mazmanian K, Guan L, Dearden L, Theil A, Moon C, Kolb B, Norian JM, Nelson, J, Wilcox J, Tan T. Trophectoderm biopsy protocols can affect clinical outcomes: time to focus on the blastocyst biopsy technique. Fertil Steril 2020; 113: 981-989.

[24] Yang D, Feng D, Gao Y, Sagnell M, Wang X, Li D. An effective method for trophectoderm biopsy using mechanical blunt dissection: a step by step demonstration. Fertil Steril 2020; 114: 438-439.

[25] Girardi L, Serdarogullari M, Patassini C, Poli M, Fabiani M, Caroselli S, Coban O, Findikli N, Boynukalin FK, Bahceci M et al. Incidence, origin, and predictive model for the detection and clinical management of segmental aneuploidies in human embryos. Am J Hum Genet 2020; 106: 525-534.

[26] Vera-Rodriguez M, Michel CE, Mercader A, Bladon AJ, Rodrigo L, Kokocinski F, Mateu E, Al-Asmar N, Blesa D, Simon C et al. Distribution patterns of segmental aneuploidies in human blastocysts identified by next-generation sequencing. Fertil Steril 2016; 105: 1047-1055.

[27] Babariya D, Fragouli E, Alfarawati S, Spath K, Wells D. The incidence and origin of segmental aneuploidy in human oocytes and preimplantation embryos. Hum Reprod 2017; 32: 2549-2560.

[28] Escriba MJ, Vendrell X, Peinado V. Segmental aneuploidy in human blastocysts: A qualitative and quantitative overview. Reprod Biol Endocrinol 2019; 17: 76.

[29] Martinez MC, Mendez C, Ferro J, Nicolas M, Serra V, Landeras J. Cytogenetic analysis of early nonviable pregnancies after assisted reproduction treatment. Fertil Steril 2010; 93: 289-292.

[30] Wellesley D, Dolk H, Boyd PA, Greenlees R, Haeusler M, Nelen V, Garne E, Khoshnood B, Doray B, Rissmann A et al. Rare chromosome abnormalities, prevalence and prenatal diagnosis rates from population-based congenital anomaly registers in Europe. Eur J Hum Genet 2012, 20: 521-526.

[31] He W, Sun X, Liu L, Li M, Jin H, Wang W-H. The Prevalence of Chromosomal Deletions Relating to Developmental Delay and/or Intellectual Disability in Human Euploid Blastocysts. PLoS ONE 2014; 9: e85207. doi:10.1371/journal.pone.0085207.

Chapter 7

Haematological Malignancies: Overview of the Recent Progresses in Genetics

Mounia Bendari, Sofia Sraidi and Nisrine Khoubila

Abstract

Genetic defects play a major role in pathogenesis of the most of haematological malignancies, including cytogenetic abnormalities, gene mutations, and abnormal gene expression. Our knowledge about the genetics of haematological disorders has been dramatically improved during the past decade, due to revolution of sequencing technologies which have played a crucial role. In this chapter, we describe the techniques commonly employed for elucidating chromosomal aberrations, prognostic impact of recurrent chromosomal abnormalities, and recently updated risk stratification systems. We will summarise the chromosomal abnormalities recently identified on many of haematological diseases such acute myeloid leukaemia, acute lymphoid leukaemia, myelodysplasic syndrome, multiple myeloma, meylopro-liferative disease and clarify their impacts on clinical phenotype and prognosis, as well as their role in the pathogenesis of these diseases. The aim of this chapter is to provide a brief overview of the recent progresses in haematological diseases genetics.

Keywords: clonal evolution, cytogenetic, somatic mutation, chromosomal abnormality, WHO classification

1. Introduction

The few recent years seen revolution on genetic technic with a big improvement of new technology. Haematological malignancy benefits from this development. Geneticists cooperate closely with hemato-oncologists; this cooperation improves many aspect of hemato-oncologists practice. It s can be helpful for diagnostic tests but also the key its allow also a better knowledge of cancer genetics which helps the specialist assess prognosis of their patient, selection of the most appropriate anticancer therapy, and monitoring the response to treatment [1].

Cytogenetic techniques occupied an important place in haematological diagnostics. The first chromosomal aberration described was in 1960 by Nowel and Hungeford characteristic, it was about CML- Philadelphia chromosome, few years after, it was be proved that almost every haematological neoplasm possess was associated with karyotype abnormalities [2].

ON 1980s, new method was developed: fluorescence in situ hybridization (FISH), it s characterised by high sensitivity and specificity, this technic has an other major particularity, it can be performed rapidly, and can classified the nature of chromosomal abnormalities [3].

Over the last few years, more powerful technologies were developed, the most remarkable one is was the Next-generation sequencing (NGS) witch was a real revolution on haematology [4].

In this chapter, we describe the techniques commonly employed for elucidating chromosomal aberrations, prognostic impact of recurrent chromosomal abnormalities, and recently updated risk stratification systems.

We will summaries the chromosomal abnormalities recently identified on many of haematological diseases such acute myeloid leukaemia, acute lymphoid leukaemia, myelodysplasic syndrome, multiple myeloma, meyloproliferative disease and clarify their impacts on clinical phenotype and prognosis, as well as their role in the pathogenesis of these diseases.

The aim of this chapter is to provide a brief overview of the recent progresses in haematological diseases genetics.

2. Cytogenetic technic

Cytogenetic study plays a crucial role on haematology, its the main outil for making diagnosis for almost all haematological malignancies, and its have an important impact prognosis for those diseases, and cytogenetic abnormalities are included on almost all prognosis score and risk stratifications for haematological neoplasms.

The past several years' remarkable efforts were deployed for better understanding of genetics and genome biology, many new technologies were developed, and the old technic saw improved their sensibility and specificity.

Veritable revolution was seen thanks to NGS, this technic offers possibility of broad analysis of a genome by whole-genome sequencing (WGS), exome sequencing, transcriptome sequencing, and epigenomics [5].

2.1 Conventional cytogenetic

Cytogenetic analyse in hematologicl neoplasms is performed by bone marrow aspiration in sterile way with heparine filled probes.

Sometimes, katyotype can be realised by peripheral blood.

The Celle of aspirated bonne marrow are cultured in vitro, then microscopic slides with metaphases chromosomes and/or interphase nuclei is performed.

Karyotype needs many metaphases cell (20 to 30) to be significatif, so its required time [2].

Conventional cytogenetics still be the most frequently ordered genetic test for various leukaemias, most prominently chronic myelogenous leukaemia (CML) in a resource limited situation.

2.2 Fluorescence in situ hybridization (FISH)

FISH is the best alternative to karyotype, it is rapid technic, with high level of specificity and sensibility. It can be realised from bone marrow or peripheral blood, it can also be performed from fixed and sectioned tissue [3].

FISH constitues a big step for studying somatic chromosomal mosaicism and molecular cytogenetic detection of chromosomal variations in interphase nuclei [6, 7].

FISH is a molecular cytogenetic technique, it identifies chromosomal abnormalities using molecular technology.

Technic of FISH is based on A DNA probe is tagged with a fluorescent marker. The probe and target DNA are denatured, and the probe is allowed to hybridise with the target. The fluorescent tag is then detected with a fluorescent microscope.

2.3 Polymerase chain reaction (PCR)

PCR is a technique that involves amplification of a desired segment of DNA by using primers, nucleotides and enzymes like reverse transcriptase and DNA polymerases. It represents the most used molecular technique on haematology. Different types of PCRs exist. Reverse transcriptase PCR (RT-PCR), Real time PCR (RQ-PCR).

In clinical practice RQ PCR is commonly used for viral copies detection and specific gene detection like Bcr-abl and PML-RARA for assessment of treatment response [1] new technique was developted: Digital PCR it s used for DNA/RNA detection and quantification. It is emerging as an alternative to conventional RQ-PCR for quantification and low abundance mutation detection [8].

PCR have many applications in hematologic malignancies include. it s used for detection of fusion genes and mutations. Its also performed for analysing of post transplant chimerism, and can be realised for determination of lymphoid clonality.

2.4 Genome-wide arrays

Microarray based testing such as array comparative genomic hybridization (CGH) and single nucleotide polymorphisms (SNP) arrays are now more used in routine diagnostics for haematological malignancies.

The copy numbers of DNA sequences in the test and reference samples are quantified by assessment of relative fluorescence intensities detected by digital imaging systems.

2.5 Gene expression profiling

This technique is based on DNA microarray which utilises plates which have various complementary genetic sequence covalently attached to them.

At present availability of GEP is limited to few research centers only limiting its wide use in clinical practice.

2.6 New generation sequencing

Over the past few years, an important increasing of the use of NGS on haematology have been shown, new platforms are available and are very helpful to identify the genetic basis of haematological neoplasms and genome biology.

Next-generation sequencing (NGS) encompasses several different methodologies that allow the investigation of genomics, transcriptomics and epigenomics [4].

Application of NGS in hematologic malignancies has confirmed presence of a lot of mutation of certain genes like TP53, ATM, RAS etc. the inconvenient for NGS, its the cost, this technic still expensive and can not be used on large spectre today especially for limited resource's country.

3. Haematological malignancies applications

Haematological neoplasms benefits from progress of biological technology, the use of new platforms helps to approve the performance of identification of genetic abnormalities.

This knowledge is crucial and have clinical utility, it can also improve diagnostics, prognosis, monitoring of minimal residual disease and it can be helpful to target of dysregulated signalling pathways by specific therapeutic targets [4].

We will summaries the chromosomal abnormalities recently identified on many of haematological diseases such acute myeloid leukaemia, acute lymphoid leukaemia, myelodysplasic syndrome, multiple myeloma, chronic lymphoblastic leukaemia, meyloproliferative disease and clarify their impacts on clinical phenotype and prognosis, as well as their role in the pathogenesis of these diseases.

3.1 Acute myeloid leukaemia (AML)

Cytogenetic abnormalities are frequently reported in the literature describing the presence of chromosomal rearrangements in important cases of acute myeloid leukaemia (AML): the rate can reach 50–60% of cases of AML [9].

It has been proved that AML is a complex and evaluative disease [10, 11]. There are many leukaemia genes, most of which are infrequently mutated, and patients typically have more than one driver mutation. The AML evolved over time, with multiple competing clones coexisting at any time [10, 11].

Over the few recent years, genome biology have seen a veritable revolution of technology, including chromosome banding, with fluorescence/chromosome in situ hybridization, or other analyses like array comparative network genomic hybridization, genome breakpoints cloning and Sanger Sequencing of candidate genes and profiling of single nucleotide polymorphism, and even whole-genome sequencing (WGS), whole-exome sequencing (WES), and RNA sequencing have all contributed to incremental improvements in understanding the genetic basis of the AML.

The whole-genome sequencing for AML showed that it is an evaluative and complex disease. There are many leukemia genes, most of which are infrequently mutated, and patients typically have many driver mutations. The evolution is characterized by emergence of many competing clones which can coexist at any time.

In fact, it has been proved that different genes and clones coexisting in the same patient, **Figure 1** illustrates that clearly [12].

200 AML patients has been analysed by The Cancer Genome Atlas (TCGA) consortium, they use whole-genome or whole-exome sequencing and they identified 23 genes as "significantly mutated" at a higher-than-expected frequency [13].

Conventional cytogenetics is very important on AML, it identifies chromosomal abnormalities, it can be balanced translocations, inversions, insertions, monosomies, and trisomies, which are present in approximately 55% of adult cases and 80% of children with AML. These are the strongest prognostic factors for response to treatment and survival in multivariate analysis. The 2008 WHO classification categorized AML based on cytogenetic or molecular abnormalities [14, 15].

The WHO 2008 and 2016 classifications incorporated modifications that allowed for a greater number of patients to be classified into the category of AML [16, 17].

Even patient with normal karyotype AML, it has been proved recently with certitude that those patients constituted very heterogeneous group; new technology helps to identified many gene mutations in normal karyotype AML by cutting-edge next-generation sequencing NGS technology, like FLT3-ITD, NPM1, CEBPA, and other additional mutations.

The most important predictors of shorter overall survival in AML patients aged less than 60 years are represented by DNMT3A and RUNX1 mutations especially those with intermediate-risk cytogenetic.

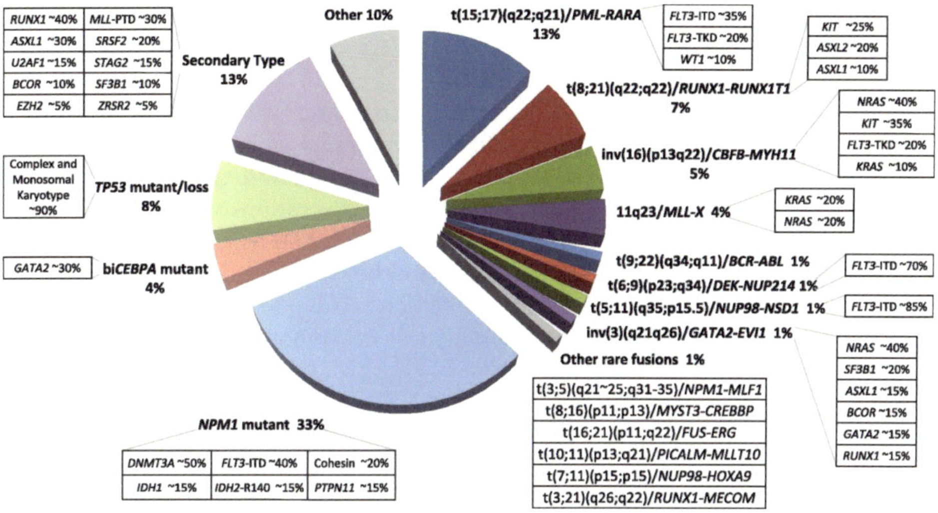

Figure 1.
Molecular classes of AML and concurrent gene mutations in adult patients up to the age of ~65 years. For each AML class denoted in the pie chart, frequent co-occurring mutations are shown in the respective boxes. Data on the frequency of genetic lesions are compiled from the databases of the British Medical Research Council (MRC) and the German-Austrian AML study group (AMLSG) and from selected studies. It indicates cohesin genes including RAD21 (10%), SMC1A (5%), and SMC3 (5%); inv(16)(p13.1q22) or t(16;16)(p13.1;q22); CBFB-MYH11; and inv(3)(q21.3q26.2) or t(3;3)(q21.3;q26.2); GATA2,MECOM(EVI1), and TP53 mutations are found in 45% and complex karyotypes in 70% of this class.

NPM1 mutations were also considered as important molecular prognosticators of Overall Survivor, particularly in the absence of FLT3-ITD, mutated TP53, and bi-allelic CEBPA mutations.

Actually, for treating patient with AML, it is indispensable to perform the research of these gene mutations. It s important for diagnosis and it can be helpful as molecular marker of prognosis, and its necessary to predictive for response of treatment, and used also for disease monitoring.

Studies demonstrated that patients with cytogenetically normal AML or intermediate-risk abnormalities have more additional gene mutations than patients with favourable or unfavourable abnormal cytogenetic and especially those with balanced translocations [18].

Recent research showed that aged patients have more driver gene mutations than younger patients. Its seems that elderly patients have more alterations in specific genes including TET2, RUNX1, ASXL1, and SRSF2. All this gene has recently been implicated in age-related clonal haematopoiesis. These found participate to improve our understanding of knowledge in AML biology between younger and older patients [19].

The application for biological technology such NSG are multiple, for exemple, there are a number of FLT3 inhibitors at various stages of clinical development were produced, such as PKC412 (midostaurin), CEP-701 (lestaurtinib), or MLN518 (tandutinib).

TKIs are promising agents in the treatment of AML patients with an FLT3-ITD mutation, especially when they are combined with chemotherapy [20].

3.2 Acute lymphoblastic leukaemia (ALL)

Acute lymphoblastic leukaemia (ALL) is the most often childhood neoplasm occuring about 30% of all cancer.

Abnormalities in chromosome number as well as structural rearrangements (translocations) are detected in 60–80% of patients with ALL, whereas the remaining 20–40% have a normal karyotype [21, 22].

Improvement of cytogenetic, FISH, and reverse transcription polymerase chain reaction (RT-PCR) analyses permits to identify subgroups of acute lymphoblastic leukaemia with specific chromosome abnormalities and allow determining treatment strategy for childhood ALL, especially when specific aberrations are present [23, 24].

Recurrent genetic abnormalities have been identified on ALL, including balanced translocations and aneuploidies. Based on the World Health Organization (WHO) classification, BCP-ALL is categorized into ALL with hyperdiploidy (>50 chromosomes), ALL with hypodiploidy (<44 chromosomes), and ALL with translocation t(9;22) (q34;q11.2) encoding BCR–ABL1, t(12;21) (p13;q22) encoding TEL–AML1, t(1;19) (q23;p13.3) encoding E2A–PBX1, t(5;14) (q31;q32) encoding IL3–IGH, and rearrangement of MLL at 11q23, with a diverse range of partner genes [25, 26].

Concerning T- ALL, common alterations include rearrangement of the T-cell receptor gene loci to transcription factor genes including TLX1, TLX3, LYL1, TAL1, and MLL [27].

ALL genomes are not static but exhibit acquisition of new chromosomal abnormalities over time. Single-nucleotide polymorphism microarray profiling studies of matched diagnosis–relapse ALL samples show that most ALL cases exhibit changes in the patterns of structural genomic alterations from diagnosis to relapse and that many relapse-acquired lesions, including those targeting genes associated with high-risk ALL (IKZF1, IKZF2, CDKN2A, and CDKN2B), are detectable at the diagnosis [28, 29].

3.3 Myelodysplasic syndrome

Myelodysplasia syndromes (MDS) are defined by a heterogeneous group of myeloid malignancies characterised by peripheral blood cytopenia and dishematopoiesis and frequently progress to acute myeloid leukaemia.

The 2016 revision defines 10 MDS subtypesas follows:

- MDS with single lineage dysplasia (MDS-SLD),
- MDS with dysplasia in two or more myeloid lineages (MDS-MLD),
- MDS-SLD/MLD with ≥15% ring sideroblasts (RSs; MDS-MLD-RS),
- MDS with an excess of blasts of up to 9% in bone marrow and up to 4% in peripheral blood (MDSEB- 1),
- MDS with 10%–19% bone marrow and 5%–19% blood blasts (MDS-EB-2),
- MDS with isolated deletion of chromosome 5q [del(5q)]
- MDS unclassifiable (MDS-U) based on defining cytogenetic abnormality, MDS-U with SLD and pancytopenia and MDS-U with 1% blood blasts.

Conventional cytogenetic allow the identification of abnormalities in approximately 50% of MDS. Some of cytogenetic abnormalities are characteristic of MDS, they may be considered as specific to MDS if the clinical context is appropriate such del(5q).

Majority of of MDS (90%) presents somatic mutations, those mutations identify molecular pathways that drive the pathogenesis of MDS. Even low abundance mutations can have prognostic value as they identify emerging clones before they impact clinical parameters.

Recent studies demonstrated that 65% of MDS patients harboured mutations in RNA splicing (SF3B1, SRSF2, U2AF1, ZRSR2) [28], 47% harbouring mutations in DNA methylation genes (DNMT3A, IDH1/2, TET2) [29, 30] and 28% in histone modification genes (ASXL1,BCOR, EZH2).

Mutations in isocitrate dehydrogenase 1 or 2 (IDH1 and IDH2) are important to identify at the time of diagnosis of high- or very high-risk MDS. These particular mutations lead to abnormal leukemogenesis. Mutated IDH1 or IDH2 are not common and are only found in approximately 4% to 12% of patients with MDS. Those gene mutations have treatment impact. Recently, two IDH inhibitors, specifically ivosidenib targeting IDH1 and enasidenib for IDH2, are approved by the United States Food and Drug Administration (FDA) for use in AML, but not in MDS [30, 31].

At present, NGS is rarely incorporated into clinical guidelines although an increasing number of studies have demonstrated the benefit of using NGS in the clinical management of MDS patients [32].

3.4 Multiple myeloma

Multiple myeloma is a malignant disease characterised by proliferation of monoclonal plasma cells leading to clinical features that include hypercalcaemia, renal dysfunction, anaemia, and bone disease (frequently referred to by the acronym CRAB) which represent evidence of end organ failure.

Recent studies have confirmed that myeloma is an heterogeneous disease composed of multiple molecularly-defined subtypes each with varying clinicopathological features and disease outcomes [33].

Chromosomal translocations account for 40–50% of primary events in myeloma and strongly influence disease phenotype [34].

Karyotypes are complex, hyperploidy can be seen in2/3 of cases, karyotypes can change from normal to abnormal during evolution of multiple myoloma.

Fluorescence in situ hybridization (FISH) ssems to be more adequate for recognising specific chromosomal changes in quiescent cells and increases the proportion of detection of chromosomal abnormalities in MM up to more than 90% [35].

IG rearrangements: translocations involving 14q32 are found in at least 65–70% of patients, most of them result from short segments exchange and are detected quite exclusively by FISH.

The (4; 14) is present in 15% of myeloma cases and has been associated with a poor prognosis in a variety of clinical settings such as those receiving high dose therapy with autologous stem cell transplant (ASCT).

The (11; 14) is observed in approximately 17% of myeloma patients and also directly up regulates a cyclin D gene in the form CCND1.

The (6; 14) is a rare translocation occurring in 2% of myeloma patients which results in the direct up regulation of the CCND3 gene [36].

Other translocations with IGH involving are reported, but they are rare and it seems that they are secondary [37].

Like other haematological neoplasm, multiple myelom benefits from the development of molecular technic like NGS, the knowledge about pathogenesis and the progression of disease has been improved, with apparition of a new concept called subclonality. In fact, NGS characterised the the wide molecular heterogeneity of the disease and the frequent occurrence of some supposedly "driver" mutations only in subclones. Those found are important for the targeted future therapies [38].

3.5 Meyloproliferative neoplasm (MPN)

Myeloproliferative neoplasms (MPN) can be denifed by a group of diseases characterised by increased proliferation of erythroid, megakaryocytic, or granulocytic.

According to WHO (2008) MPN regroups clonal disorders of myeloid progenitor cells., MPN have been classified into 3 groups. Also called Philadelphia chromosome-negative (Ph –), myeloproliferative neoplasms (MPNs), include polycythemia vera (PV), essential thrombocythemia (ET), and myelofibrosis (both primary (PMF) and secondary (post-ET/PV MF)) [39].

The 3 MPN entities are characterised by their clinical heterogeinicity, establishment of precise diagnosis at disease onset can be a real challeng for physician. The World Health Organisation (WHO) established revision for diagnostic and defined new criteria for MPN on 2016, (**Figure 2**) [40].

The first gene mutation described in 2005, JAK2-V617F, turned out to be the most important and most frequently recurring somatic mutation in MPN [6–9]. The frequency ofJAK2-V617F is around 95% in PV and between 50% and 60% in ET and PMF [41].

MPN patients who do not present mutations in any of the aforementioned genes (so-named "triple-negative" MPN cases), but those patients seem to have hyperactive JAK2 signalling [42].

The *JAK2*V617F mutation arises in a multipotent haematopoietic progenitor, is present in all myeloid lineages.

The *JAK2*V617F is mainly restricted to classical MPNs with the exception of refractory anaemia with ring sideroblasts and thrombocytosis (RARS-T). It can rarely be found in some other malignant hemopathies [43, 44].

Concerning MPL mutation, it have been demonstrated that two types of mutation exist: (*MPL*; the thrombopoietin [TPO] receptor [TPOR]) mutations located in exon 10, both have been reported on association with MPNs.

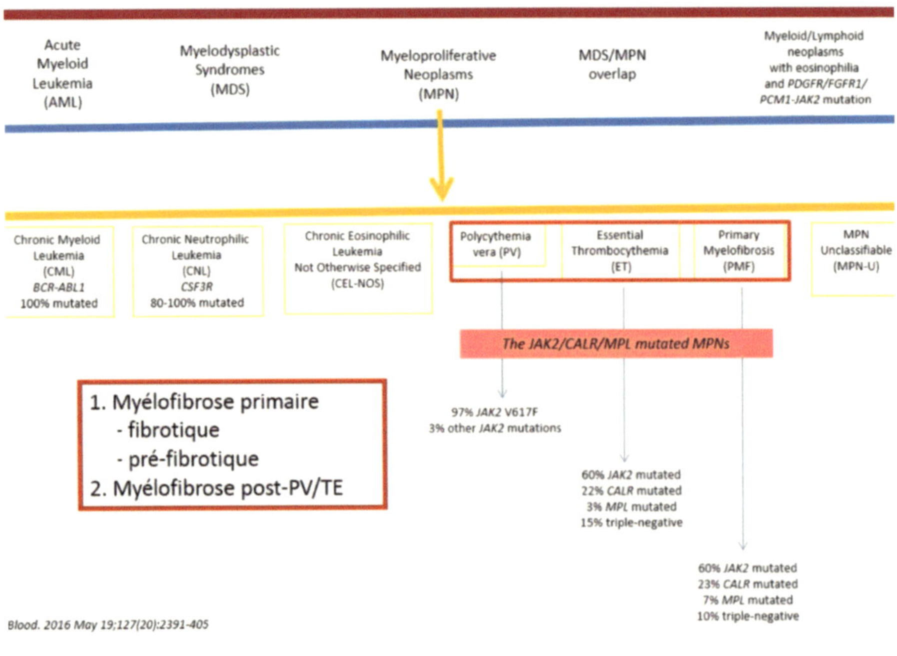

Figure 2.
2016 WHO classification of myeloid malignancies.

The most frequent are mutations on the tryptophan W515 located at the boundary of the transmembrane and the cytosolic domains of MPL, the most prominent mutations being MPLW515L and K [42].

Mutations on *MPL*W515 are restricted to ET (around 3%) and PMF (around 5%).

At the end of 2013, frameshift mutations in the *CALR* gene were identified in the majority of *JAK2*⁻ and *MPL*⁻ ET and PMF (50%–60% ET and 75% PMF).

There are great differences in the frequency between type 1 and type 2 mutations in ET and PMF.

CALR mutations reported are often heterozygous, only few cases of homozygous mutations have been reported, more particularly for type 2 mutations [43].

It has been also proved that CALR mutants gives a stronger clonal advantage when compared with *JAK2*V617F.

In 2010, somatic mutations in exon 2 of *LNK* (*SH2B3*), an adaptor protein which regulates JAK2 activation, were detected in 2 patients (PMF and ET).

The 3 MPN oncogenes are considered as true drivers of the disease phenotype with JAK2 exon 12 giving only an erythrocytosis phenotype, JAK2V617F giving rise to ET, PV, and MF, whereas CALR mutant and MPLW515L/K/A are associated with ET and MF, resembling the phenotype observed in patients.

The 3 main driver mutations do not explain the entire heterogeneity of the classical *BCR-ABL*⁻ MPNs. The 3 main driver mutations do not explain the entire heterogeneity of the classical *BCR-ABL*⁻ MPNs.

Like other hematologic malignancies, the mutations in epigenetic regulators can be observed, such as MDS and AML; and some of the gene mutations, such as TET2 and ASXL1, are more frequent in MDS than in MPN.

Some studies suggest that the presence of mutations in TET2, EZH2, and ASXL1 are associated with hight risk of secondary AML.

Few time after the discovery of the JAK2 V617F mutation, multiple small molecule inhibitors were developed for therapeutic use: ruxolitinib, is the first JAK1 and JAK2 inhibitor, approved in August 2011 for use in intermediate and high-risk PMF and post PV/ET myelofibrosis.

Selective JAK2 inhibitors, (SAR302503 and BMS911543), combination JAK2/JAK3 inhibitor (CEP701), and combination JAK2/TYK2 inhibitor (pacritinib) have shown clinical efficacy in phase I/II trials.

4. Conclusion

Major progress has been achieved in understanding the molecular pathogenesis of haematological malignancies in a very short period of 10 years.

While the clinical utility of this genetic and epigenetic revolution, novel therapeutic agents aimed at the aberrant underlying processes are more and more included rational combination therapies. This knowledge increase outcome for haematological diseases patients, and are helpful to develop therapies based on insights into the genetic basis of these haematological neoplasms.

Author details

Mounia Bendari[1]*, Sofia Sraidi[2] and Nisrine Khoubila[2]

1 Mohammed VI University of Health Sciences (UM6SS), Casablanca, Morocco

2 Faculty of Medicine and Pharmacy of Casablanca, Hassan II University, Casablanca, Morocco

*Address all correspondence to: bendarimounia@gmail.com

References

[1] E. Mały. Cytogenetics in Hematooncology Recent Trends in Cytogenetic Studies–Methodologies 58 and Applications

[2] Alessandro Gozzetti, michelle M le beau. Fluorescence in situ hybridization: Uses and limitations November 2000 Seminars in Hematology 37(4):320-333

[3] Esteban braggio Lessons from next-generation sequencing analysis in hematological malignancies Blood Cancer Journal 3(7):e127

[4] Kobold DC, Steinberg KM, Larson DE, Wilson RK, Mardis ER. The next generation sequencing revolution and its impact on genomics. Cell 2013; 155:27-38

[5] Yurov YB, Iourov IY, Vorsanova SG, et al (2007) Aneuploidy and confi ned chromosomal mosaicism in the developing human brain. PLoS One 2(6):e558. D

[6] Yurov YB, Vorsanova SG, Liehr T, et al (2014) X chromosome aneuploidy in the Alzheimer's disease brain. Mol Cytogenet 7(1):20.

[7] M. Bendari and al, Current Cytogenetic Abnormalities in Acute Myeloid Leukemia. DOI: http://dx.doi.org/10.5772/intechopen.91425

[8] Grimwade D, Hills RK, Moorman AV, Walker H, Chatters S, Goldstone AH, et al. Refinement of cytogenetic classification in acute myeloid leukemia: Determination of prognostic significance of rare recurring chromosomal abnormalities among 5876 younger adult patients treated In the United Kingdom Medical Research Council trials. Blood. 2010;116(3):354-365

[9] Grimwade D, Walker H, Oliver F, et al. On behalf of the Medical Research Council adult and children's leukaemia working parties. The importance of diagnostic Cytogenetics on outcome in AML: Analysis of 1612 patients entered into the MRC AML10 trial. Blood.1998;92(7):2322-2333

[10] Döhner H, Estey E, Grimwade D, Amadori S, Appelbaum FR, Büchner T, et al. Diagnosis and management of AML in adults: 2017 ELN recommendations from an international expert panel. Blood. 2017;129(4)

[11] Cortes JE, Kantarjian H, Foran JM, et al. Phase I study of quizartinib administered daily to patients with relapsed or refractory acute myeloid leukemia irrespective of FMS-like tyrosine kinase 3-internal tandem duplication status. Journal of Clinical Oncology. 2013;31: 3681-3687

[12] Galanis A, Ma H, Rajkhowa T, et al. Crenolanib is a potent inhibitor of FLT3 with activity against resistanceconferring point mutants. Blood. 2014;123:94-100

[13] Metzelder SK, Schroeder T, Finck A, et al. High activity of Sorafenib in FLT3-ITD-positive acute myeloid leukemia synergizes with allo-immune effects to induce sustained responses. Leukemia. 2012;26:2353-2359

[14] Campo E, and al, The 2008 WHO classification of lymphoid neoplasms and beyond: evolving concepts and practical applications Blood. 2011 May 12; 117(19): 5019-5032.

[15] Dohner H. Implications of the molecular characterization of acute myeloid leukemia. Hematology 2007:412-417.

[16] Peterson L, Bloomfield CD, Döhner H, Niemeyer C, Godley L. Myeloid neoplasms with germline predisposition.

[17] Swerdlow S, Campo E, Harris NL, et al., editors. World Health Organization Classification of Tumours of Haematopoietic and Lymphoid Tissues. 4th ed. Lyon, France

[18] Raimondi SC, Pui CH, Head DR, Rivera GK, Behm FG. Cytogenetically different leukemic clones at relapse of childhood acute lymphoblastic leukemia. Blood 1993;82:576-580.

[19] Yang JJ, Bhojwani D, Yang W, et al. Genome-wide copy number profiling reveals molecular evolution from diagnosis to relapse in childhood acute lymphoblastic leukemia. Blood 2008;112:4178-4183.

[20] Russell LJ, Capasso M, Vater I, et al. Deregulated expression of cytokine receptor gene, CRLF2, is involved in lymphoid transformation in B-cell precursor acute lymphoblastic leukemia. Blood 2009;114:2688-2698.

[21] Mullighan CG, Collins-Underwood JR, Phillips LA, et al. Rearrangement of CRLF2 in B-progenitor- and Down syndrome-associated acute lymphoblastic leukemia. Nat Genet 2009;41:1243-1246.

[22] Swerdlow SH, Campo E, Harris NL, Jaffe ES, Pileri SA, Stein H, et al., editors. WHO classification of tumours of haematopoietic and lymphoid tissues. Lyon: IARC; 2008.

[23] Deschler B and Lübbert M: Acute myeloid leukemia: Epidemiology and etiology. Cancer 107: 2099-2107, 2006.

[24] Armstrong SA and Look AT: Molecular genetics of acute lymphoblastic leukemia. J Clin Oncol 23: 6306-6315, 2005.

[25] Harrison CJ (2001) Detection and significance of chromosomal abnormalities in childhood acute lymphoblastic leukemia. Blood Rev 15:49-59 doi:10.1054/blre.2001.0150

[26] Kearney L (1999) The impact of new FISH technologies on the cytogenetics of the haematological malignancies. Br J Haematol 104: 648-658 doi:10.104 6/j.1365-2141.1999.01181

[27] Graubert TA, Shen D, Ding L, Okeyo-Owuor T, Lunn CL, Shao J, et al. Recurrent mutations in the U2AF1 splicing factor in myelodysplastic syndromes. Nat Genet. 2011;44:53-57.

[28] Delhommeau F, Dupont S, Valle VD, James C, Trannoy S, Massé A, et al. Mutation in TET2 in myeloid cancers. New Engl J Med. 2009;360:2289-2301.

[29] Mardis ER, Ding L, Dooling DJ, Larson DE, McLellan MD, Chen K, et al. Recurring mutations found by sequencing an acute myeloid leukemia genome. N Engl J Med. 2009; 361:1058-1066.

[30] Gil-Perez A, Montalban-Bravo G. Management of myelodysplastic syndromes after failure or response to hypomethylating agents. Ther Adv Hematol. 2019; 10:2040620719847059.

[31] DiNardo CD, Jabbour E, Ravandi F, et al. IDH1 and IDH2 mutations in myelodysplastic syndromes and role in disease progression. Leukemia. 2016;30(4):980-984.

[32] Tobiasson M A. O. Kittang3Treatment of myelodysplastic syndrome in the era of next-generation sequencing M. 2019 The Association for the Publication of the Journal of Internal Medicine

[33] Steven M. Prideaux, Emma Conway O'Brien, and Timothy J. Chevassut. The Genetic Architecture of Multiple Myeloma. Advances in Hematology 2014(6):864058

[34] W. M. Kuehl and P. L. Bergsagel, "Early genetic events provide the basis for a clinical classification of multiple

myeloma," Hematology, pp. 346-352, 2005.

[35] F. Zhan, Y. Huang, S. Colla et al., "The molecular classification of multiple myeloma," Blood, vol. 108, no. 6, pp. 2020-2028, 2006.

[36] Avet-Loiseau H, Attal M, Moreau P, et al. Genetic abnormalities and survival in multiple myeloma: the experience of the Intergroupe Francophone du Myelome. Blood. 2007;109(8):3489-3495.

[37] J. Corre. Genetics of multiple myeloma: another heterogeneity level? Blood (2015) 125 (12) : 1870-1876.

[38] James C, Ugo V, Le Couedic JP, et al. A unique clonal JAK2 mutation leading to constitutive signalling causes polycythaemia vera. Nature. 2005;434:1144-1148.

[39] Baxter EJ, Scott LM, Campbell PJ, et al. Acquired mutation of the tyrosine kinase JAK2 in human myeloproliferative disorders. Lancet. 2005;365:1054-1061

[40] Arber DA, Orazi A, Hasserjian R, et al. The 2016 revision to the World Health Organization classification of myeloid neoplasms and acute leukemia. Blood. 2016;127(20):2391-2405

[41] Rampal R, Al-Shahrour F, Abdel-Wahab O, et al. Integrated genomic analysis illustrates the central role of JAK-STAT pathway activation in myeloproliferative neoplasm pathogenesis. Blood. 2014;123: e123–e133.

[42] Levine RL, Loriaux M, Huntly BJ, et al. The JAK2V617F activating mutation occurs in chronic myelomonocytic leukemia and acute myeloid leukemia, but not in acute lymphoblastic leukemia or chronic lymphocytic leukemia. Blood. 2005;106 (10):3377-3379.

[43] Pikman Y, Lee B, Mercher T, et al. MPLW515L is a novel somatic activating mutation in myelofibrosis with myeloid metaplasi. PLoS Med. 2006:3 (7):e270.

[44] Klampfl T, Gisslinger H, Harutyunyan AS, et al. Somatic mutations of calreticulin in myeloproliferative neoplasms. N Engl J Med. 2013 ;369 (25):2379-2390

Chapter 8

Genetic Abnormalities in ALL

Mounia Bendari, Sofia Sraidi and Nisrine Khoubila

Abstract

Acute lymphoblastic leukemia (ALL), can be defined by a family of genetically heterogeneous lymphoid neoplasms derived from B- and T-lymphoid progenitors. ALL constitutes the most common childhood cancer, due to an overproduction of immature lymphoid hematopoietic cells. Genetic analyzes currently provides important information for classifying patients into prognostic groups, genetic analysis also helps to understand the mechanisms of relapse, pharmacogenetics and the development of new potential therapeutic targets, which should help to further improve the results of leukemia. In fact, the new techniques in molecular cytogenetic permits to identify new cryptic abnormalities, these discoveries have led to the development of new therapeutic protocols. The role of cytogenetic analysis is crucial on ALL patient's management. Karyotyping coupled with FISH analysis identifies recurrent chromosomal abnormalities in ALL, many of these abnormalities have prognostic and treatment impact. This chapter summarizes chromosomal abnormalities that are common and classify ALL according to the World Health Organization (WHO) classifications (2016 revision). We will present the main genetic modifications recently identified as well as the sequence mutations which have helped in the elucidation of the pathogenesis of ALL.

Keywords: Acute lymphoblastic leukemia, World Health Organization classification, cytogenetic analysis, FISH analysis

1. Introduction

Acute lymphoblastic leukemias (ALL) are clonal proliferations of immature cells involved in B (LAL-B) or T (LAL-T) lymphoid differentiation and blocked at an early stage of differentiation. The ALL is the most frequent childhood malignancy. In multiple studies dating back more than 50 years, both B-cell ALL and T-cell ALL are associated with characteristic and recurrent cytogenetic changes [1, 2]. They had a great value for diagnosis, risk stratification, disease monitoring and treatment selection. The conventional cytogenetics techniques have experienced significant advancement into molecular cytogenetics technologies. These recent advancements have largely overcome the limitations of conventional cytogenetics techniques. Fluorescence in situ hybridization (FISH), multiplex ligation-dependent probe amplification (MLPA), array comparative genomic hybridization (aCGH) and next-generation sequencing (NGS) techniques are part of the armory of molecular cytogenetics technologies [3–5].

2. Cytogenetic technics

2.1 Conventional cytogenetic

Conventional banded karyotyping for the detection and prognosis of genetic diagnosis is considered as the gold standard. It has been used to analyze genome modifications that include both genome gains and losses, as well as rearrangements within and between chromosomes [5]. Conventional single cell and metaphase cytogenetics are important in tumor genetics for disease control, tumor staging, and research purposes to recognize chromosomal regions containing genes and proto-oncogeneses of putative tumor suppressors [6].

2.2 Fluorescence in situ hybridization (FISH)

Fluorescence in situ hybridization (FISH) is a technique for determining complex DNA sequences as well as the number and structure of chromosomes. The method is focused on the use of fluorescent probes that can recognize specific DNA sequences. FISH is a technique for detecting genetic defects in embryos that is fast and sensitive. Targeting and denaturing DNA fixed in cells, nuclei, or metaphase chromosomes on the surface of the slide is the basis of the FISH analysis. Next, after its denaturation, a complementary single-stranded DNA sequence probe will precisely re-anneal double-stranded DNA (hybrid) molecules during the hybridization reaction. Probe DNA molecules are labeled enzymatically with modified nucleotides. They are DNA molecules designated hapten-labeled (indirect FISH) and fluorescent-labeled (direct FISH). An antifade solution containing 4′,6-diamidino-2 phenylindole is added to the slide after the removal of unbound single-stranded DNA and nonspecifically bound DNA from the slide by posthybridization washing. Using epifluorescence microscopes with specialized filters for detecting fluorochromes, FISH signals are observed. The signal is captured by a charge-coupled system camera, and the fluorescent signals are then quantified [7, 8].

2.3 Polymerase chain reaction (PCR)

Polymerase chain reaction (PCR) is a fast and cheap technique used to "amplify" small segments of DNA by copying them. Because significant amounts of a sample of DNA are needed for molecular and genetic analysis, without PCR amplification, studies of isolated pieces of DNA are almost impossible. The PCR method is based on a cell's natural processes for replicating a new DNA strand. For PCR, only a few biological ingredients are needed. The template DNA—that is, the DNA that contains the region to be copied, such as a gene—is an essential component. A prototype can be as small as one DNA molecule. The sequence of two short regions of nucleotides (DNA subunits) at either end of the region of interest is all that is needed for this fragment to be replicated. The primers bind to the template at their complementary sites, or anneal, and serve as the starting point for copying. The replication of the desired intervening sequence is achieved when DNA synthesis at one primer is guided toward the other. Free nucleotides and a DNA polymerase, an enzyme that builds new DNA strands by sequentially adding on free nucleotides according to the template's instructions, are also needed.

2.4 Next-generation sequencing (NGS)

NGS requires sequencing of millions of DNA molecules concurrently to produce sequence reads. In order to detect small insertions/deletions (indels) and structural variants (SVs) of 450 bp, sequence reads are aligned with the reference genome and base variants. Overall, NGS has the potential to generate up to one billion short reads per instrument cycle, an immense amount of data cheaply.

3. Chromosomal and molecular abnormalities associated with ALL-B

60–80% of patients with ALL have abnormalities in chromosome number or structural rearrangements (translocations), whereas the remaining 20–40% have normal karyotype [9, 10]. Besides those with a normal karyotype, t(9;22)(q34,q11); BCR/ABL (BCR-ABL1), t(12;21)(p13;q22);TEL/AML1 (ETV6-RUNX1), t(4;11)(q21;q23); MLL/AF4 (KMT2A/AFF1), t(1;19)(q23;p13); E2A/PBX1 (TCF3-PBX1), are the most common cytogenetic subtypes in ALL [10–12].

3.1 Structural chromosomal abnormalities

3.1.1 The t(9;22)(q34;q11.2)(BCR-ABL1)

The BCR/ABL1 or Philadelphia (Ph) chromosome is a t(9;22) product that fuses the chromosome 9 Abelson kinase gene (ABL1) with the chromosome 22 breakpoint cluster region (BCR) that expresses the fusion protein BCR-ABL1: a constitutively active tyrosine kinase. The breakpoint occurs between exons 1 and 2 (e1 and e2) of the BCR gene in the minor breakpoint cluster region, m-BCR, in the majority of Ph positive ALL patients, and between exons 1 and 2 of the ABL gene in the majority of Ph positive ALL patients (e1a2). This results in the development of a 7-kb mRNA and the expression of the p190 protein [13]. This transcribes an 8.5 kb mRNA that codes for a chimeric p210 protein [14]. The Philadelphia chromosome is the most important cytogenetic abnormality. It is seen in 3% of pediatric patients, and almost 25% of adults, and rises with age, reflecting about half of the cases of patients older than 60 years of age, and although historically associated with poor prognosis, results have been markedly improved with the use of tyrosine kinase inhibitors (TKIs) [15].

3.1.2 The t(12;21)(p13;q22)(ETV6-RUNX1)

The most prevalent translocation in childhood acute lymphoblastic leukemia is TEL-AML1 gene fusion, induced by t(12;21) (p12;q22). However, this anomaly is rare among adults. The translocation of t(12;21) is cryptic by normal G-banding and includes FISH examination for cytogenetic detection [16]. ETV6-RUNX1 patients were thought to have a good prognosis at first, and they were associated with favorable risk factors including female gender, young age, low white cell count, and CD10+ immunophenotype [17]. However, some studies found no gain for ETV6-RUNX1 patients [18], while others found a high incidence of gene fusion in relapse patients and a predilection for late relapse [19, 20]. However, it is now clear that the initial optimism was justified. Almost every major clinical trial group in the world has confirmed that children with the ETV6-RUNX1 fusion have excellent overall survival and very low relapse rates [21, 22], and the presence of added cytogenetic or molecular abnormalities does not modify this good prognosis.

3.1.3 The KMT2A (MLL) Gene Rearrangements (11q23)

A transcriptional coactivator with methyltransferase activity encodes the gene KMT2A. The rearrangements result in the fusion of the 5′ portion of KMT2A, including the methyltransferase domain, to the 3′ region of the partner genes.

KMT2A (MLL) rearrangements, particularly the translocation of t(4;11)(q21;q23), are most common in infants (<1 year of age) and are associated with poor outcomes [23, 24]. The t(4;11)(q21;q23) can be detected by conventional cytogenetics, FISH, RT-PCR, or Southern blot techniques. Overall, ALL with MLL rearrangement have an unfavorable prognosis.

3.1.4 The t(1;19)(q23;P13), TCF3-PBX1

The translocation t(1;19)(q23;p13) is the result of the fusion of the PBX1 gene at 1q23 with the TCF3 (E2A) gene at 19p13.3. This translocation occurs in approximately 5–6% of childhood and adult B-cell precursor (BCP) ALLs [25, 26]. The translocation t(1.19) appears in balanced form (presence of two derived chromosome) or more often in unbalanced form with the derivative chromosome 19: (der (19)t(1.19)(q23;p13)). TCF3-PBX1 patients usually have a pre-B immunophenotype that expresses cytoplasmic μ [27]. It's one of the few genetic disorders that doesn't seem to increase in frequency with age. Originally considered a high-risk subtype of ALL, it is now associated with a favorable outcome with contemporary treatment, although some studies have indicated that it has an independent risk factor for CNS relapse [28].

3.1.5 IKZF1(7p12) deletion or mutations

The IKZF1 gene is located on the 7p12.2 chromosome band, consists of 8 exons, and codes for the transcription factor IKAROS with key regulatory functions in lymphopoiesis [29, 30]. IKAROS harbors 6 fingers zinc. Four of these are located in the DNA-binding domain encoded by exons 4 to 6 and are important for the tumor suppressor function of IKAROS to be preserved. Exon 8 encodes the remaining 2 zinc fingers and mediates IKAROS dimerization either as a homodimer or with other transcription factors in its family [29, 31]. The deletions of this gene, are very frequently associated with the BCR-ABL1 fusion in the development of ALL of the B line. These deletions result in haploinsufficiency by partial or total deletion. it seems that these deletions represent an independent risk of relapse.

3.1.6 CRLF2 rearrangement (IGH-CRLF2; P2RY8-CRLF2)

CRLF2 encodes cytokine receptor-like factor 2, also known as the thymic stromal-derived lymphopoietin receptor (TSLPR), wich forms an heterodimeric receptor with the interleukin-7 receptor a chain (IL7Ra) for thymic stromal lymphopoietin (TSLP). CRLF2 is deregulated by translocation into the immunoglobulin heavy chain locus (IGH-CRLF2); focal deletion upstream of CRLF2, resulting in P2RY8-CRLF2 fusion; and less often, CRLF2 point mutations (F232C) [32]. In Ph-like and Down syndrome-related ALL, CRLF2 rearrangements are most common and are age dependent, with P2RY8-CRLF2 associated with young age and IGH-CRLF2 associated with older age and Hispanic ancestry [33, 34]. Most CRLF2-rearranged ALLs have additional JAK–STAT or Ras signaling alterations, particularly activating JAK1 or JAK2 mutations, FLT3 and IL7R sequence mutations, SH2B3 deletions, TSLP rearrangements, and Ras mutations [35–37]. CRLF2 rearrangements have been associated with poor prognosis in most studies, especially in cases of concomitant IKZF1 alteration [38, 39].

3.1.7 Intrachromosomal amplification of chromosome 21 (iAMP21)

Intrachromosomal amplification of chromosome 21 or iAMP21 is defined as the presence of three or more additional copies of *RUNX1* on a structurally abnormal chromosome 21. The iAMP21 chromosome is often initially detected by *ETV6 - RUNX1* FISH analysis [40, 41]. It affects about 2–5% of B-cell precursor acute lymphoblastic leukemia pediatric patients [42, 43]. Patients with iAMP21 are usually between the ages of 7 and 13, with a median age of 10 [44]. It is particularly uncommon in children under the age of five and in people over the age of twenty. Complex intrachromosomal amplification of chromosome 21 is most common in older children and the poor prognosis is improved by high-risk treatment. Accurate identification of this abnormality is considered to be extremely necessary in determining the best course of treatment.

3.1.8 The Philadelphia Chromosome – like Acute Lymphoblastic Leukemia (Ph-like ALL)

Ph-like, or BCR-ABL1-like ALL is characterized by a leukemic cell transcriptional profile similar to Ph + ALL but lack the BCR-ABL1 fusion gene [45, 46]. Ph-like ALL is vary heterogeneous in the altered genes and the form (rearrangements, mutations, or deletions) of alterations that result in the activated tyrosine kinase or cytokine receptor signaling characteristic of this subtype of ALL [46]. However, these fall into four main groups: (1) Alterations driving JAK–STAT signaling, most commonly rearrangements of CRLF2 (IGH-CRLF2, P2RY8-CRLF2, CRLF2 F232C), and less commonly, rearrangements of JAK2, EPOR, or TYK2, and mutations/deletions of IL7R, SH2B3, JAK1, JAK3, TYK2, IL2RB); (2) fusions involving ABL-class genes (ABL1, ABL2, CSF1R, LYN, PDGFRA, PDGFRB); (3) mutations activating Ras signaling (NRAS, KRAS, PTPN11); and (4) less common fusions (FLT3, FGFR1, NTRK3) [35, 36, 47]. Ph-like is associated with high-risk clinical characteristics, poor response to induction chemotherapy, elevated levels of minimal residual disease (MRD), and/or poor survival [48].

3.2 ALL with number anomalies

3.2.1 Hyperdiploidy

Hyperdiploidy is the most prevalent recurrent abnormality in childhood B-ALL. In the World Health Organization classification of tumors of hematopoietic and lymphoid tissues, hyperdiploidy in B-lymphoblastic leukemia (B-ALL), characterized by the presence of 51–65 chromosomes, has been identified as a distinct subtype of B-ALL [49]. In hyperdiploidy, numerical chromosomal gains are non-random, with additional copies (usually trisomies) of chromosomes 21, X, 14, and 4 most commonly found in pediatric patients [50]. Despite the presence of non-specific structural abnormalities, the extra chromosomes are still normal copies of chromosomes. There is a poor understanding of the mechanism involved in inducing hyperdiploidy and its role in leukaemogenesis. Hyperdiploid B-ALL comprises approximately 25–30% of pediatric B-ALL cases [51]; and is often associated with a favorable prognosis with a cure rate greater than 90%, especially when hyperdiploidy is associated with trisomies of chromosomes 4 and 10 [52–54].

3.2.2 Hypodiploidy

Hypodiploidy, characterized by less than 44 chromosomes is less frequent than hyperdiploid ALL. Three cytogenetic subgroups of hypodiploidy were defined: near haploidy, with 24–31 chromosomes; low hypodiploid, with 32–39 chromosomes; and high hypodiploid, with 40–43 chromosomes [55]. Near-haploidy patients showed common chromosomal gains, rare structural abnormalities and a co-incident doubled hyperdiploid population [56–58]. Low hypodiploidy karyotypes are usually monosomic for chromosomes 3, 7, 15, 16, 17, and disomic for chromosomes 1, 6, 11, and 18. In this subgroup, the phenomenon of doubling-up occurs, and sub-clones with near-triploid karyotypes are common. Furthermore, evidence indicates that in near-haploid situations, cytogenetic research is more likely to show only the doubled-up clone. Overall Hypodiploid acute lymphoblastic leukemia (ALL) has been associated with a dismal prognosis [59, 60].

4. Chromosomal and molecular abnormalities associated with ALL-T

T-cell acute lymphoblastic leukemia (T-ALL) is a leukemia that develops when there is an accumulation of genomic lesions that impair T-cell growth. T-ALL is correlated with a lot of genetic diversity. The accumulation of a variety of genetic and epigenetic defects leads to leukemic transformation [61]. As a result of excessive neoplastic cell proliferation, they cause disorders of cell differentiation, apoptosis, oncogene activation, and suppressor inhibition. The first genetic abnormalities in T-ALL patients were chromosome aberrations. Except for tetraploidy, which occurs in around 5% of cases, numerical changes are uncommon and have little prognostic significance.

The identification of chromosomal anomalies, such as 9p deletions that result in CDKN2A (p16) and CDKN2B (p15) inactivation, and translocations affecting T-cell receptor genes, has been crucial in gaining an understanding of the genetic defects present in T-ALL.

The proportion of cytogenetically normal cases at diagnosis is higher in T-ALL than in B-ALL, with about 50 percent of patients with T-ALL possessing a normal karyotype. Approximately one-third of T-ALL patients have a translocation involving one of the T-cell receptor genes (TCR), with a breakpoint at 14q11 (*TCRA/TCRD*) or 7q34 (*TCRB*), juxtaposing the T-cell receptor genes to pivotal transcription factor genes, such as TAL1, TAL2, LYL1, OLIG2, LMO1, LMO2, TLX1 (HOX11), TLX3 (HOX11L2), NKX2–1, NKX2–2, NKX2–5, HOXA genes, MYC, and MYB. In the adult population, the translocation t(10;14)(q24;q11.2), which results in overexpression of the *TLX1 (HOX11)* gene, is the most common and is associated with a favorable outcome [62, 63]. In addition, T-ALLs can contain cryptic rearrangements of ABL1 that may be amenable to TKI therapy. In general, studies of gene expression profiling have helped to classify T-ALL into molecular subgroups characterized by distinct signatures of gene expression and aberrant activation of specific oncogenes of the T-ALL transcription factor, including MEF2C, HOXA, TLX1, NKX2.1, TLX3, TAL1, LMO1, and LMO2 [41, 64].

5. Conclusion

Acute lymphoblastic leukaemia (ALL) is the commonest childhood cancer. However, conventional cytogenetic and molecular analyses fail to identify clonal driver alterations in approximately 25% of ALL in children and the majority of cases in adults but when they are present, they have a crucial role in the management of ALL patients. Recent advancements in gene expression profiling and genome-wide sequencing have revolutionized our understanding of ALL pathogenesis over the last years. As defined in this review, the accumulation of results has restructured ALL genetic classifications. Overall, we expect that research over the next decade can thoroughly define the genomic of ALL across all generations and refine the therapeutic algorithm to be more targeted and individualized.

Author details

Mounia Bendari[1*], Sofia Sraidi[2] and Nisrine Khoubila[2]

1 Mohammed VI University of Health Sciences, Casablanca, Morocco

2 Faculty of Medicine and Pharmacy of Casablanca, Hassan II University, Morocco

*Address all correspondence to: bendarimounia@gmail.com

References

[1] Bloomfi eld CD, Lindquist LL, Arthur D et al (1981) Chromosomal abnormalities in acute lymphoblastic leukemia. Cancer Res 41(11 Pt 2):4838-4843

[2] Harrison CJ (2009) Cytogenetics of paediatric and adolescent acute lymphoblastic leukaemia. Br J Haematol 144(2) :147-156. doi: 10.1111/j.1365-2141.2008.07417.x

[3] Gouas L, Goumy C, Ve´rone` se L, Tchirkov A, Vago P (2008). Gene dosage methods as diagnostic tools for the identification of chromosome abnormalities. Pathol Biol (Paris) 56:345-353

[4] Gijsbers AC, Ruivenkamp CA (2011). Molecular karyotyping: from microscope to SNP arrays. Horm Res Paediatr 76:208-213.

[5] Le Scouarnec S, Gribble SM (2012). Characterising chromosome rearrangements: recent technical advances in molecular cytogenetics. Heredity (Edinb) 108:75-85.

[6] George P, Wilhelm J, and Phillip B: Molecular Diagnostics; 2016 p. 249-267.

[7] Haaf T (2000). Fluorescence in situ hybridization. In: Meyers RA, editor. Encyclopedia of analytical chemistry. Chichesater, UK: John Wiley and Sons Ltd. p. 4984.

[8] Riegel M (2014). Human molecular cytogenetics: from cells to nucleotides. Genet Mol Biol 37 (Suppl):194-209.

[9] Deschler B and Lübbert M: Acute myeloid leukemia: Epidemiology and etiology. Cancer 107: 2099-2107, 2006.

[10] Armstrong SA and Look AT: Molecular genetics of acute lymphoblastic leukemia. J Clin Oncol 23: 6306-6315, 2005.

[11] Look AT: Oncogenic transcription factors in the human acute leukemias. Science 278: 1059-1064, 1997.

[12] Mullighan, C. G. (2019). How advanced are we in targeting novel subtypes of ALL? Best Practice & Research Clinical Haematology, 101095. doi:10.1016/j.beha.2019.101095

[13] Clark SS, McLaughlin J, Crist WM, *et al* Unique forms of the *abl* tyrosine kinase distinguish Ph1-positive CML from Ph1-positive ALL. *Science*, 239: 775-777, 1988

[14] Groffen J and Heisterkamp N. The BCR/ABL hybrid gene. Baillieres Clin Haematol, **1**: 983-999, 1987

[15] Iacobucci I, Mullighan CG. Genetic basis of acute lymphoblastic leukemia. J Clin Oncol. 2017;35:975-983

[16] Douet-Guilbert N, Morel F, Le Bris MJ et al (2003) A fl uorescence in situ hybridization study of TEL-AML1 fusion gene in B-cell acute lymphoblastic leukemia (1984-2001). Cancer Genet Cytogenet 144(2):143-147

[17] Shurtleff SA, Buijs A, Behm FG, Rubnitz JE, Raimondi SC, Hancock ML, et al. TEL/ AML1 fusion resulting from a cryptic t(12;21) is the most common genetic lesion in pediatric ALL and defines a subgroup of patients with an excellent prognosis. Leukemia 1995;9:1985-9.

[18] Hann I, Vora A, Harrison G, Harrison C, Martineau M, Moorman AV, et al. Determinants of outcome after intensified therapy of childhood lymphoblastic leukaemia: results from Medical Research Council United Kingdom acute lymphoblastic leukaemia XI protocol. Br J Haematol 2001;113:103-14.

[19] Harbott J, Viehmann S, Borkhardt A, Henze G, Lampert F.

Incidence of TEL/AML1 fusion gene analyzed consecutively in children with acute lymphoblastic leukemia in relapse. Blood 1997;90:4933-7.

[20] Forestier E, Heyman M, Andersen MK, Autio K, Blennow E, Borgstrom G, et al. Outcome of ETV6/RUNX1-positive childhood acute lymphoblastic leukaemia in the NOPHO-ALL-1992 protocol: frequent late relapses but good overall survival. Br J Haematol 2008;140:665-72.

[21] Moorman AV, Ensor HM, Richards SM, Chilton L, Schwab C, Kinsey SE, et al. Prognostic effect of chromosomal abnormalities in childhood B-cell precursor acute lymphoblastic leukaemia: results from the UK Medical Research Council ALL97/99 randomised trial. Lancet Oncol 2010;11:429-38

[22] Rubnitz JE, Wichlan D, Devidas M, Shuster J, Linda SB, Kurtzberg J, et al. Prospective analysis of TEL gene rearrangements in childhood acute lymphoblastic leukemia: a Children's Oncology Group study. J Clin Oncol 2008;26:2186-91.

[23] Meyer C, Hofmann J, Burmeister T, et al: The MLL recombinome of acute leukemias in 2013. Leukemia 27:2165-2176, 2013

[24] Andersson AK, Ma J, Wang J, et al: The landscape of somatic mutations in infant MLL-rearranged acute lymphoblastic leukemias. Nat Genet 47: 330-337, 2015

[25] Barber KE, Harrison CJ, Broadfield ZJ, et al: Molecular cytogenetic characterization of TCF3 (E2A)/19p13.3 rearrangements in B-cell precursor acute lymphoblastic leukemia. Genes Chromosomes Cancer 46:478-486, 2007

[26] Burmeister T, Gokbuget N, Schwartz S, et al: ¨ Clinical features and prognostic implications of TCF3- PBX1 and ETV6-RUNX1 in adult acute lymphoblastic leukemia. Haematologica 95:241-246, 2010

[27] Hunger SP. Chromosomal translocations involving the E2A gene in acute lymphoblastic leukemia: clinical features and molecular pathogenesis. Blood 1996;87:1211-24.

[28] Jeha S, Pei D, Raimondi SC, et al: Increased risk for CNS relapse in pre-B cell leukemia with the t(1;19)/TCF3-PBX1. Leukemia 23:1406-1409, 2009

[29] Georgopoulos K, Bigby M, Wang JH, et al. The Ikaros gene is required for the development of all lymphoid lineages. Cell. 1994; 79(1):143-156.

[30] Molnar A, Wu P, Largespada DA, et al. The ´ Ikaros gene encodes a family of lymphocyterestricted zinc finger DNA binding proteins, highly conserved in human and mouse. J Immunol. 1996;156(2):585-592

[31] Rebollo A, Schmitt C. Ikaros, Aiolos and Helios: transcription regulators and lymphoid malignancies. Immunol Cell Biol. 2003;81(3): 171-175

[32] Yoda A, Yoda Y, Chiaretti S, et al: Functional screening identifies CRLF2 in precursor B-cell acute lymphoblastic leukemia. Proc Natl Acad Sci USA 107: 252-257, 2010

[33] Russell LJ, Capasso M, Vater I, et al: Deregulated expression of cytokine receptor gene, CRLF2, is involved in lymphoid transformation in B-cell precursor acute lymphoblastic leukemia. Blood 114:2688-2698, 2009

[34] Mullighan CG, Collins-Underwood JR, Phillips LA, et al: Rearrangement of CRLF2 in B-progenitorand Down syndrome-associated acute lymphoblastic leukemia. Nat Genet 41:1243-1246, 2009

[35] Roberts KG, Li Y, Payne-Turner D, Harvey RC, Yang YL, Pei D, et al. Targetable kinase-activating lesions in Ph-like acute lymphoblastic leukemia. N Engl J Med 2014;371:1005-1015.

[36] Roberts KG, Gu Z, Payne-Turner D, McCastlain K, Harvey RC, Chen IM, et al. High frequency and poor outcome of Philadelphia chromosome-like acute lymphoblastic leukemia in adults. J Clin Oncol 2017;35:394-401.

[37] Mullighan CG, Zhang J, Harvey RC, et al: JAK mutations in high-risk childhood acute lymphoblastic leukemia. Proc Natl Acad Sci USA 106:9414-9418, 2009

[38] Harvey RC, Mullighan CG, Chen IM, et al: Rearrangement of CRLF2 is associated with mutation of JAK kinases, alteration of IKZF1, Hispanic/Latino ethnicity, and a poor outcome in pediatric B-progenitor acute lymphoblastic leukemia. Blood 115:5312-5321, 2010

[39] Cario G, Zimmermann M, Romey R, et al: Presence of the P2RY8-CRLF2 rearrangement is associated with a poor prognosis in non-high-risk precursor B-cell acute lymphoblastic leukemia in children treated according to the ALL-BFM 2000 protocol. Blood 115:5393-5397, 2010

[40] Busson-Le Coniat M, Nguyen Khac F, Daniel MT et al (2001) Chromosome 21 abnormalities with AML1 amplifi cation in acute lymphoblastic leukemia. Genes Chromosomes Cancer 32(3):244-249

[41] Harewood L, Robinson H, Harris R et al (2003) Amplification of AML1 on a duplicated chromosome 21 in acute lymphoblastic leukemia: a study of 20 cases. Leukemia 17(3):547-553. doi: 10.1038/sj.leu.2402849

[42] Moorman AV. The clinical relevance of chromosomal and genomic abnormalities in B-cell precursor acute lymphoblastic leukaemia. *Blood Rev.* 2012;26(3):123-135. [PubMed] [Google Scholar]

[43] Ma SK, Wan TS, Cheuk AT, Fung LF, Chan GC, Chan SY, et al. Characterization of additional genetic events in childhood acute lymphoblastic leukemia with TEL/AML1 gene fusion: a molecular cytogentics study. *Leukemia.* 2001;15(9):1442-1447. [PubMed] [Google Scholar]

[44] Moorman AV, Richards SM, Robinson HM, Strefford JC, Gibson BE, Kinsey SE, et al. Prognosis of children with acute lymphoblastic leukaemia (ALL) and intrachromosomal amplification of chromosome 21 (iAMP21). Blood 2007;109: 2327-30

[45] Den Boer ML, van Slegtenhorst M, De Menezes RX, Cheok MH, Buijs-Gladdines JG, Peters ST, et al. A subtype of childhood acute lymphoblastic leukaemia with poor treatment outcome: a genomewide classification study. Lancet Oncol 2009;10:125-134.

[46] Mullighan CG, Su X, Zhang J, Radtke I, Phillips LA, Miller CB, et al. Deletion of IKZF1 and prognosis in acute lymphoblastic leukemia. N Engl J Med 2009;360:470-480

[47] Reshmi SC, Harvey RC, Roberts KG, Stonerock E, Smith A, Jenkins H, et al. Targetable kinase gene fusions in high-risk B-ALL: a study from the Children's Oncology Group. Blood 2017;129:3352-3361.

[48] Roberts KG, Pei D, Campana D, et al: Outcomes of children with BCR-ABL1–like acute lymphoblastic leukemia treated with risk-directed therapy based on the levels of minimal residual disease. J Clin Oncol 32:3012-3020, 2014

[49] Borowitz M, Chan J, Downing J, LeBeau M, Arber D. B-lymphoblastic

Leukemia/Lymphoma with Recurrent Genetic Abnormalities. In: Swerdlow S, Campo E, Harris N, et al., eds. WHO Classification of Tumours of Haematopoietic and Lymphoid Tissues, Revised, 4th edn. Lyon: International Agency for Research on Cancer (IARC); 2017.

[50] Heerema NA, Raimondi SC, Anderson JR, et al. Specific extra chromosomes occur in a modal number dependent pattern in pediatric acute lymphoblastic leukemia. Genes Chromosomes Cancer. 2007;46:684-693

[51] Paulsson K, Johansson B. High hyperdiploid childhood acute lymphoblastic leukemia. Genes Chromosomes Cancer. 2009;48:637-660. [PubMed] [Google Scholar]

[52] Paulsson K, Forestier E, Andersen MK, et al. High modal number and triple trisomies are highly correlated favorable factors in childhood B-cell precursor high hyperdiploid acute lymphoblastic leukemia treated according to the NOPHO ALL 1992/2000 protocols. Haematologica. 2013;98:1424-1432. [PMC free article] [PubMed] [Google Scholar]

[53] Sharathkumar A, DeCamillo D, Bhambhani K, et al. Children with hyperdiploid but not triple trisomy (+4,+10,+17) acute lymphoblastic leukemia have an increased incidence of extramedullary relapse on current therapies: a single institution experience. Am J Hematol. 2008;83:34-40. [PubMed] [Google Scholar]

[54] Paulsson K, Lilljebjorn H, Biloglav A, et al. The genomic landscape of high hyperdiploid childhood acute lymphoblastic leukemia. Nat Genet. 2015;47:672-676.

[55] Harrison CJ, Moorman AV, Broadfi eld ZJ et al (2004) Three distinct subgroups of hypodiploidy in acute lymphoblastic leukaemia. Br J Haematol 125(5):552-559. doi: 10.1111/j.1365-2141.2004.04948.x

[56] Gibbons, B., MacCallum, P., Watts, E., Rohatiner, A.Z.S., Webb, D., Katz, F.E., Secker-Walker, L.M., Temperley, I.J., Harrison, C.J., Campbell, R.H.A., Nash, R., Broadbent, V. & Chessells, J.M. (1991) Near haploid acute lymphoblastic leukemia: seven new cases and a review of the literature. Leukemia, 5, 738-743.

[57] Mitelman, F., Johansson, B. & Mertens, F. (2003) Mitelman Database of Chromosome Aberrations in Cancer. (WWW document). URL

[58] Raimondi, S.C., Zhou, Y., Mathew, S., Shurtleff, S.A., Sandlund, J.T., Rivera, G.K., Behm, F.G. & Pui, C.H. (2003) Reassessment of the prognostic significance of hypodiploidy in pediatric patients with acute lymphoblastic leukemia. Cancer, 98, 2715-2722

[59] Pui CH, Williams DL, Raimondi SC, et al. Hypodiploidy is associated with a poor prognosis in childhood acute lymphoblastic leukemia. Blood. 1987;70(1): 247-253

[60] Raimondi SC, Zhou Y, Mathew S, et al. Reassessment of the prognostic significance of hypodiploidy in pediatric patients with acute lymphoblastic leukemia. Cancer. 2003;98(12):2715-2722

[61] Vicente, C.; Schwab, C.; Broux, M.; Geerdens, E.; Degryse, S.; Demeyer, S.; Lahortiga, I.; Elliott, A.; Chilton, L.; La Starza, R.; et al.Targeted sequencing identifies associations between IL7R-JAK mutations and epigenetic modulators in T-cell acute lymphoblastic leukemia. Haematologica 2015,100, 1301-1310

[62] The Groupe Français de Cytogénétique Hématologique. Cytogenetic abnormalities in adult acute lymphoblastic leukemia: correlations

with hematologic findings outcome. A Collaborative Study of the Group Français de Cytogénétique Hématologique. Blood. 1996;87(8):3135-3142.

[63] Wetzler M, Dodge RK, Mrózek K, et al. Prospective karyotype analysis in adult acute lymphoblastic leukemia: the cancer and leukemia Group B experience. Blood. 1999;93(11): 3983-3993.

[64] Ferrando AA, Neuberg DS, Staunton J, et al: Gene expression signatures define novel oncogenic pathways in T cell acute lymphoblastic leukemia. Cancer Cell 1:75-87, 2002

Chapter 9

Cytogenetic and Genetic Advances in Myelodysplasia Syndromes

Mounia Bendari and Nisrine Khoubila

Abstract

Myelodysplasia syndromes (MDS) are defined by a heterogeneous group of myeloid malignancies characterized by peripheral blood cytopenia and dishematopoiesis and frequently progress to acute myeloid leukemia. Conventional karyotype has a crucial role in myelodysplastic syndrome (MDS) and is one of items of the International Prognostic Scoring System (IPSS) for patient risk stratification and treatment selection. Approximately 50–60% of cases of MDS present chromosomal abnormalities, like the deletions of chromosome 5q and 7q, trisomy 8, and complex karyotypes. New genomic technologies have been developted, like single-nucleotide polymorphism array and next-generation sequencing. They can identify the heterozygous deletions wich result in haplo-insufficient gene expression (e.g., CSNK1A1, DDX41 on chromosome 5, CUX1, LUC7L2, EZH2 on chromosome 7) involved in the pathogenesis of myelodysplasia syndromes. Genetic abnormalities are multiple, the most recurrent one are involved in the RNA splicing like SF3B1, SRSF2, U2AF1, ZRSR2, LUC7L2, and DDX41. Epigenetic modifications are also identified, such as histone modification as ASXL1, EZH2. Finally, it can be DNA methylation (e.g., TET2, DNMT3A, IDH1/IDH2). On this review we will summarize the most recent progress in molecular pathogenesis of MDS, and try to better understand the pathogenesis of the specific subgroups of MDS patients and applications of discovery of new genetic mutation in the development of new therapeutic.

Keywords: cytogenetic, new genomic technologies, IPSS-R, karyotype, myelodysplasia, single-nucleotide polymorphism array, next-generation sequencing

1. Introduction

Myelodysplastic syndromes (MDSs) comprises a heterogeneous group of myeloid neoplasms, they are characterized by pancytopenia, bone marrow (BM) hyperplasia, dysplasia, and cytopenias of the peripheral blood. The blast count may be normal or elevated but is less than 20% in the bone marrow and peripheral blood. MDS are characterized by elevated risk of progression to secondary acute myeloid leukemia (AML) [1, 2].

MDS is caused by accumulation of genetic or epigenetic (such as promoter hypermethyltion) lesions, first it s occurred in an immature progenitor and provides proliferative advantage of the MDS clone over normal immature progenitors.

MDS progenitors leads abnormal terminal differentiation and capacity to resist to apoptosis. These two features explain the clinical consequences of blast accumulation and peripheral cytopenias. Microenvironmental changes and immune deregulation participate to this differentiation defect [3].

The purpose of this review is to overview the recent advances in the cytogenetics and genetics of MDS and related disorders.

2. Epidemiology

MDS's incidence increase markedly with age, and the classical patient will be in their late 60s or 70s and have one or more otherwise unexplained cytopenia [4, 5].

The incidence is 4 to 5 per 100,000 persons per year, the direct etiology for MDS is usually unknown. However, in 15 to 20% of cases, MDS are secondary (sMDS) to chemotherapy and/or radiotherapy for an other disease. Some times, MDS can be, secondary to exposure to benzene or other aromatic hydrocarbons, or products used in agriculture.

The pathophysiology of MDS and its progression to AML involve cytogenetic, genetic, and epigenetic factors [6].

Now, it is well recognized that MDS is, like other cancers, shaped by recursive rounds of positive selections, where gene mutations and other genetic alterations play central roles.

3. WHO Classification of MDS

The classification of Tumours of Haematopoetic and Lymphoid Tissues done by the World Health Organization (WHO) defines MDS as a clonal, stem cell disorder.

The 2016 new revision of this classification defines ten MDS subtypes. The first subtype is defined by MDS with single lineage dysplasia (MDS-SLD), the second one is characterized by MDS with dysplasia in two or more myeloid lineages (MDS-MLD). Third subtype is MDS-SLD/MLD with ≥15% ring sideroblasts (RSs; MDS-MLD-RS). An excess of blasts of up to 9% in bone marrow and up to 4% in peripheral blood define MDSEB- 1, and MDS with 10–19% bone marrow and 5–19% blood blasts define MDS-EB-2. An other subtype is present on this classification, it s MDS with isolated deletion of chromosome 5q [del(5q)]. Finally, we found MDS unclassifiable (MDS-U) based on defining cytogenetic abnormality, MDS-U with SLD and pancytopenia and MDS-U with 1% blood blasts [7].

Table 1 summarizes 2016 OMS classification of MDS.

4. Evolution/prognostic

Until 2016, del(5q) was the only genetic marker implicated in MDS classification. In the updated classification, identification of SF3B1 mutation determines MDS-RS (even when the RS count is >5–15%).

The revised International Prognostic Scoring System (IPSS-R) for MDS propose 5 risk groups depending on number and severity of cytopenia. Its include also the percentage of bone marrow blasts and cytogenetic aberrations **Figure 1** [7].

Cytogenetic abnormalities were categorized into 5 prognostic subgroups that were shown to have significant prognostic relevance with different median survival and risk of evolution into AML.

The molecular profile of MDS has become a vital factor in assessing the risk of patients with MDS and making treatment decisions. Health care providers must understand when to order genetic testing, how to interpret the results, and the

Name	Dysplastic lineages	Cytopenias*	Ring sideroblasts as % of marrow erythroid elements	BM and PB blasts	Cytogenetics by conventional karyotype analysis
MDS with single lineage dysplasia (MDS-SLD)	1	1 or2	<15%/<5%†	BM <5%, PB <1%, no Auer rods	Any, unless fulfills all criteria for MDS with isolated del(5q)
MDS with multilineage dysplasia (MDS-MLD)	2 or 3	1–3	<15%/<5%†	BM <5%, PB <1%, no Auer rods	Any, unless fulfills all criteria for MDS with isolated del(5q)
MDS with ring sideroblasts (MDS-RS)					
MDS-RS with single lineage dysplasia (MDS-RS-SLD)	1	1 or2	≥15%/≥5%†	BM <5%, PB <1%, no Auer rods	Any, unless fulfills all criteria for MDS with isolated del(5q)
MDS-RS with multllineage dysplasia (MDS-RS-MLD)	2 or 3	1–3	≥15%/≥5%†	BM <5%, PB <1%, no Auer rods	Any, unless fulfills all criteria for MDS with isolated del(5q)
MDS with Isolated del(5q)	1–3	1–2	None or any	BM <5%, PB <1%, no Auer rods	del(5q) alone or with 1 additional abnormality except −7 or del (7q)
MDS with excess blasts (MDS-EB)					
MDS-EB-1	0–3	1–3	None or any	BM 5–9% or PB 2–4%, no Auer rods	Any
MDS-EB-2	0–3	1–3	None or any	BM 10–19% or PB 5–19% or Auer rods	Any

Name	Dysplastic lineages	Cytopenias*	Ring sideroblasts as % of marrow erythroid elements	BM and PB blasts	Cytogenetics by conventional karyotype analysis
MDS, unclassifiable (MDS-U)					
with 1% blood blasts	1–3	1–3	None or any	BM <5%, PB – 1%,‡ no Auer rods	Any
with single lineage dysplasia and pancytopenia	1	3	None or any	BM <5%, PB <1%, no Auer rods	Any
based on defining cytogenetic abnormality	0	1–3	<15%	BM <5%, PB <1%, no Auer rods	MDS-defining abnormality
Refractory cytopenia of childhood	1–3	1–3	None	BM <5%, PB <2%	Any

**Cytopenias defined as: hemoglobin, <10 g/dL; platelet count, <100 × 10^9/L; and absolute neutrophil count, <1.8 × 10^9/L. Rarely, MDS may present with mild anemia or thrombocytopenia above these levels. PB monocytes must be <1 × 10^9/L.*
†If SF3B1 mutation is present.
‡One percent PB blasts must be recorded on at least 2 separate occasions.

Table 1.
The World Health Organization (WHO) classification of Tumours of Haematopoetic and Lymphoid Tissues defines MDS.

Prognostic Variable	0	0.5	1.0	1.5	2.0	3	4
Cytogenetics*	Very good	—	Good	—	Intermediate	Poor	Very poor
BM blasts (%)	≤2	—	>2% – <5%	—	5%–10%	>10%	—
Hemoglobin	≥10	—	8 – <10	<8	—	—	—
Platelets	≥100	50 – <100	<50	—	—	—	—
ANC	≥0.8	<0.8	—	—	—	—	—

Scores for risk groups are as follows: Very low ≤1.5; Low >1.5–3; Intermediate >3–4.5; High >4.5–6; Very high >6.

ANC = absolute neutrophil count; BM = bone marrow.

*Cytogenetics: Very good: −Y, del(11q); Good: normal, del(5q), del(12p), del(20q), double including del(5q); Intermediate: del(7q), +8, +19, i(17q), any other single or double independent clones; Poor: −7, inv(3)/t(3q)/del(3q), double including −7/del(7q), complex: 3 abnormalities; Very poor: complex: >3 abnormalities.

Reprinted with permission from Greenberg PL, Tuechler H, Schanz J, et al. Revised international prognostic scoring system for myelodysplastic syndromes. Blood 2012;120:2454–65.

Figure 1.
International Prognostic Scoring System (IPSS).

Prognostic subgroups, % of patients	Cytogenetic abnormalities	Median survival,* y	Median AML evolution, 25%,* y	Hazard ratios OS/AML*	Hazard ratios OS/AML†
Very good (4%*/3%†)	−Y, del(11q)	5.4	NR	0.7/0.4	0.5/0.5
Good (72%*/66%†)	Normal, del(5q), del(12p), del(20q), double including del(5q)	4.8	9.4	1/1	1/1
Intermediate (13%*/19%†)	del(7q), +8, +19, i(17q), any other single or double independent clones	2.7	2.5	1.5/1.8	1.6/2.2
Poor (4%*/5%†)	−7, inv(3)/t(3q)/del(3q), double including −7/del(7q), complex: 3 abnormalities	1.5	1.7	2.3/2.3	2.6/3.4
Very poor (7%*/7%†)	Complex: > 3 abnormalities	0.7	0.7	3.8/3.6	4.2/4.9

OS indicates overall survival; and NR, not reached.
*Data from patients in this IWG-PM database, multivariate analysis (n = 7012).
†Data from Schanz et al[8] (n = 2754).

Figure 2.
Revised International Prognostic Scoring System (IPSS-R).

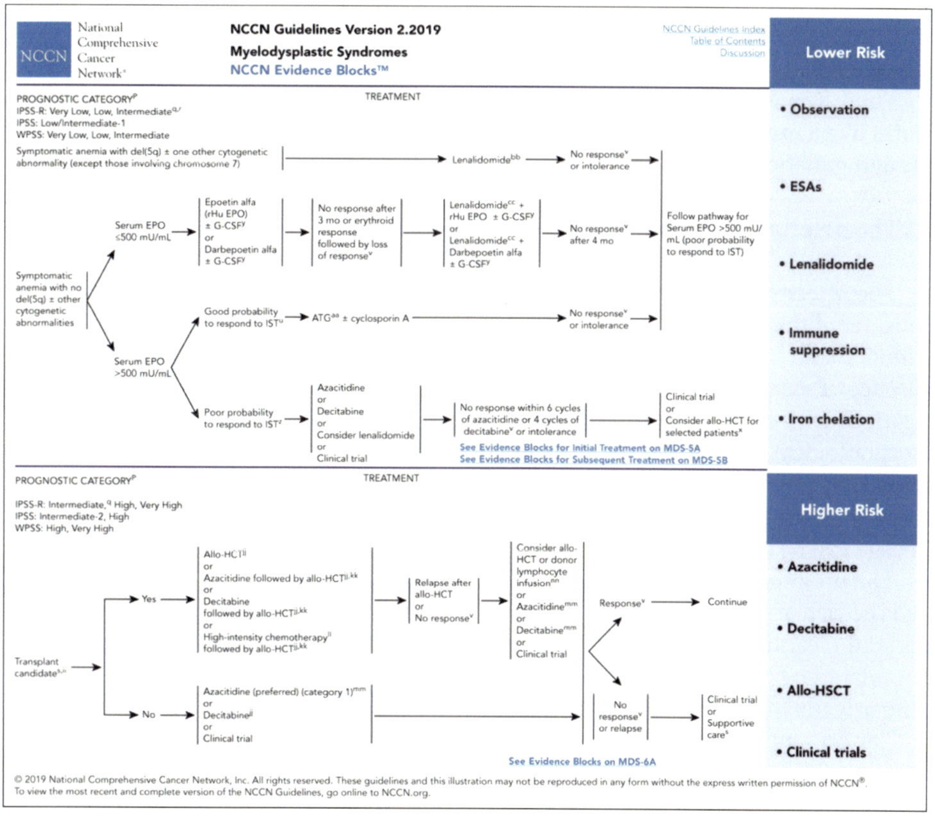

Figure 3.
The National Comprehensive Cancer Network (NCCN) guidelines stratify patients on low and high risk patient with different treatment options.

implications of specific cytogenetic abnormalities on the Revised International Prognostic Scoring System (IPSS-R). In addition, the genetic profile of a patient's disease may dictate which therapy is most appropriate (**Figure 2**).

In fact, risk stratification in the first step in the care of newly diagnosed MDS is crucial, it help to convey disease severity, can set expectations (overall survivor for months, years, or decade) and is important to define treatment strategy.

The National Comprehensive Cancer Network (NCCN) guidelines stratify patients on low and high risk patient with different treatment options (**Figure 3**).

5. Cytogenetics

Conventional prognostic scoring of MDS is based on the extent of cytopenia, the percentage of bone marrow blast infiltration, and karyotype abnormalities [8, 9].

Metaphase cytogenetic show presence of abnormalities in approximately 50% of MDS. Some of cytogenetic abnormalities are characteristic of MDS, they may be considered as specific to MDS if the clinical context is appropriate.

Only del(5q) may be considered as MDS subtype. Therefore, the 2016 revision of the WHO classification considered cases with del(5q) plus one other abnormality to be categorized as MDS with isolated del(5q), providing the second abnormality is not del(7q).

Acute myeloid leukemia (AML) is characterised by balanced abnormalities wich predominate, in contrast of MDS where unbalanced abnormalities are more common. Overall, the most frequent abnormalities are loss of the Y chromosome (-Y), del(5q), +8, del(20q), and – 7 [10, 11].

On the other hand, cytogenetic abnormalities are more frequent in therapy-related MDS (t-MDS) than de novo MDS, being reported in 70–90% cases [12].

Thus, a constitutional karyotype on a blood sample cultivated using phytohemaglutin like a mitogen can be realised in these two cases for a right interpretation of the cytogenetic response after treatment.

5.1 The deletion of 5q and "5q- syndrome"

The interstitial deletion of the long arm of chromosome 5 (del (5q)), can be considered the most frequent cytogenetic aberration in MDS, it occurs in 15% of patients with MDS. The "5q syndrome" is defined by an isolated del (5q) and an absence of excess blast, on the peripheral blood or the marrow. The IPSS includes a patient with del (5q) isolated in a favorable group. These patients have distinct morphological characteristics, thrombocytosis is found in one third of patients, macrocytic anemia and hypolobulated megakaryocytes, on the other hand, showed little dysplasia along the granulomonocytic and erythroid lines. The erythroid lineage can be hypoplastic.

The prognosis is favorable, overall survival for patients with 5q- is prolonged and the risk of progression to AML is lower than other patients with MDS. Even without treatment, clonal evolution is rarely reported. The prognosis is indeed dominated by the consequences of chronic transfusions, but these patients respond dramatically to the immunomodulating agent lenalidomide [13].

The 5q suppression can vary in size, but invariably affects bands q31 to q33. A common 5q33 deletion spanning more than 1.5 Mb and encompassing 42 genes has been reported. A model of haploinsufficiency in which the loss of a single copy of one or more genes possibly responsible for the 5q syndrome is suggested, and this may be explained by the absence of recurrent point mutation or cryptic deletion on the allele 5q normal.

5.2 Trisomy 8

Trisomy 8 + 8, (10–15%) that sometimes results from germinal mosaicism, is often subclonal, fluctuating independently of blast counts.

5.3 Monosomy 7/deletion 7q

Chromosome 7 anomalies comports del(7q), monsomy 7 (−7/del(7q)), or more rarely t(1;7), are second in frequencies after del(5q) (10%) and they have a poor Prognostic value on overall survival and risk of transformation.

Prognostic values can be possibly distinct according to regions of deletion. In fact, many and different minimal regions of deletion have been noted in 7q35–36.

Monosomy 7 can change constitutional bone marrow failures syndromes (FA, Down syndrome) or AA, or arise after radiation or toxic exposure. It is the most frequent alteration in childhood MDS where it is often associated with a degree of myeloproliferation.

Now it s known that G-CSF treatment may select a − 7 clone, and that 7q is a genetically instable region.

Patients with −7/del(7q) are characterized by neutrophil functional impairment, they are exposed to severe infections. Those patient is very have poor response to intensive chemotherapy but respond better to hypomethylating agents [14, 15].

5.4 3q26 abnormalities

The IPSS-R considers the 3q26 alterations: inv.(3)(q21q26), and t(3;3)(q21;q26) as pejorative abnormalities, they rearrange the MECOM (MDS1/EVI1) locus with complex oncogenic roles and may be accompanied by thrombocytosis. Numerous other partners of EVI1 are reported as PRDM16 in t(1;3)(p36;q21) and RUNX1 in t(3;21)(q26;q22).

5.5 17p-/-17 and TP53 mutations

Abnormalities of chromosome 17 are multiple, it can be deletion, monosomy, unbalanced translocation or isochromosome 17 which involve the loss of one TP53 locus, they are described in sMDS/AML after treatment with chemotherapy and/or radiotherapy, usually its associate with other complex genetic abnormalities.

It has been proved that chromosome 17p deletion with consistent involvement of TP53 gene located at 17p13 is associated with vacuolated pseudo-Pelger-Huet granulocytes. Those patients have poor prognosis both in MDS and AML.

5.6 Complex karyotypes

Karyotypes is considered as Complex (15%) by the presence of at least 3 anomalies that are thought to result from alterations in DNA repair or checkpoint signalling. Complex karyotypes are usully heterogeneous. The prognosis of patient having complex karyotype is worsening with each additional aberration, rather than by the chromosomes involved (most frequently, 5, 7 and 17). Complex karyotypes are by essence chemoresistant, but interesting results with the hypomethylating agent decitabine have been described by some studies.

5.7 Others abnormalities

According to the IPSS, other abnormalities can be considered as favourable:

- del(20q) which is not considered as specific of MDS, but has been related to a particularly presentation involving frequent thrombocytopenia.

- Recurrent unbalanced translocations involving 1q have been identified in primary MDS with a partial trisomy for the long arm of chromosome 1: t(1;15)(q11;p11); t(Y;1)(q12;q12), der(16) t(1;16)(q11;q11).

- secondary MDS are charactherized by translocation associated with the long arm of chromosome 7, we can also found: Deletion 9q, del(11q), del(12p) and del(13q) witch are recurrent in MDS.

- deletion of short arm of chromosome 12, del(12p) are variable. It can be associated with multiple karyotypic changes in sMDS. De novo disorders are rare with and 12p- chromosome as a sole aberration is rarely seen. Deletions are interstitial, with loss of material between band p11 and p13.

- Acquired monosomy X can be sporadically found in female MDS patients. Xq13 may also be involved in translocations in MDS, as well as in rearrangements such as an isodicentric chromosome X with breakpoint at q13 (idic (X)(q13))

Thus, chromosomal aberrations still have clinical relevance in MDS even in the era of genomic medicine. Because they basically consist in copy number changes, their detection will likely be improved by array-based karyotyping and/or by massive parallel sequencing itself [16].

6. Genetics

Novel genomic tools are now available that can both confirm clonality and provide valuable prognostic information (**Figure 4**).

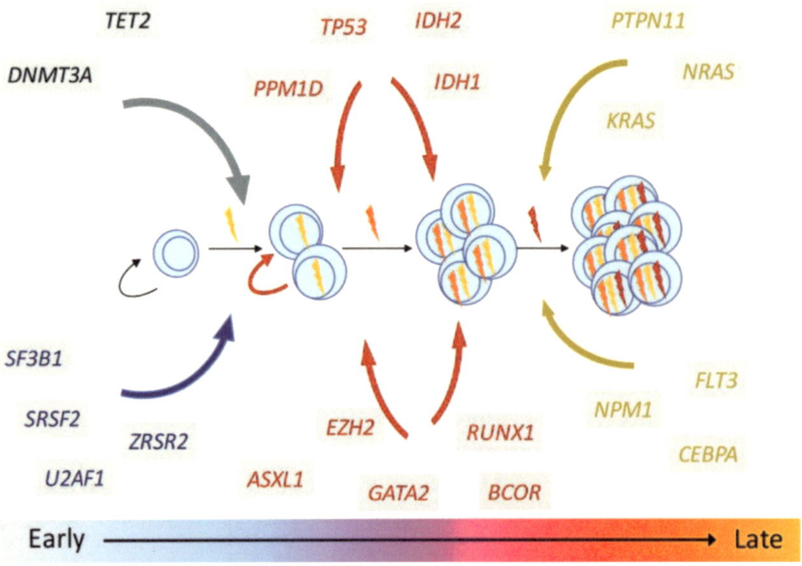

Figure 4.
Most frequently mutated pathways: serial acquisition of somaticmutations causes clonal stem cell expansion and impaired differentiation.

It is crucial to the understanding of the pathogenesis and disease phenotypes of MDS to decipher the mutations that are involved in the positive selection ("driver" mutations) and the mechanisms by which those mutations are positively selected.

Next-generation sequencing (NGS) has revealed a landscape of genetic alterations including coding exons and copy number alterations [17, 18].

The most common genetic alterations in MDS are mutations affecting RNA splicing and epigenetic modifier pathways. Those mutations are found on MDS more than AML, They are implicated on pathogenesis of MDS rather than primary AML.

Many insights have been done about the implication of these mutations on RNA splicing and the epigenome, and initial murine models of several of these mutations have been reported [19].

Mutations are frequently associated with specific disease phenotype, drug response, and clinical outcomes, and thus, it is essential to be familiar with MDS genetics for better management of patients.

MDS is typically driven by a multistep genetic process with recurrent mutations affecting basic cellular pathways, including RNA splicing, epigenome regulation, and myeloid transcriptional coordination, those abnormalities caused DNA damage and provoked stress responses, and growth factor signalling.

MDS is characterised by a lot of recurrent mutation genes and diversify of affected pathways. However, myeloid driver mutations have common fundamental biological property: they all can be responsible of clonal dominance at the stem cell level.

The diversity of clinical MDS phenotypes associated with specific mutations may be attributable to differential correlation of the hematopoietic stem cell HSC self-renewal program and lineage-specific differentiation programs.

More than 90% of MDS have somatic mutations, those mutations identify molecular pathways that drive the pathogenesis of MDS. Even low abundance mutations can have prognostic value as they identify emerging clones before they impact clinical parameters.

Among major mutational targets in MDS are the molecules involved in DNA methylations, chromatin modification, RNA splicing, transcription, signal transduction, cohesion regulation, and DNA repair.

NGS using whole-exome sequencing showed that MDS patients carry a median of 9 somatic mutations in the entire coding region, those mutations include driver mutations that advance clonal selection and passenger mutations (random mutations) that do not promote disease [20].

If we focus on the most reccurent mutated pathways, 65% of MDS patients harboured mutations in RNA splicing (SF3B1, SRSF2, U2AF1, ZRSR2) [12], followed by 47% harbouring mutations touching DNA methylation genes (DNMT3A, IDH1/2, TET2) and 28% in histone modification genes (ASXL1,BCOR, EZH2) **Figure 3** [21–23].

MDS and primary AML share common mutational targets, pleading for the same pathogenesis in different neoplasms. However, the recurrence of these mutations differed between MDS and primary AML; in MDS overrepresentation of mutations in splicing factors (SFs) and epigenetic regulators are often reported, in contrast of AML, genetic abnormalities include mutations in receptor tyrosine kinases like FLT3, RAS pathway genes, and CEBPA and IDH1/2, witch are the most frequent mutations reported [24].

Recurrent mutations are described in genes regulating DNA methylation (DNMT3A, TET2, IDH1/2), and post-translational chromatin modification (EZH2, ASXL1). Also transcription regulation (TP53, RUNX1, GATA2), are found, such as the RNA spliceosome machinery (SF3B1, U2AF1, SRSF2, ZRSR2), cohesion complexes (STAG2), and signal transduction (JAK2, KRAS, CBL) [21].

Mutations in TP53, EZH2, ETV6, RUNX1, SRSF2 and ASXL1 occurs low survivals. [24] These mutations can predict responses treatment by hypomethylating agents and allogeneic HSCT.

Furthermore, internal tandem duplication of FLT3 (FLT3 -ITD), have been described during MDS progression and represent potential therapeutic targets [25, 26].

Therefore, a better knowledge of the molecular landscape in MDS has crucial role for determination of implications on treatment response, prognostication, and novel molecular therapeutic targeting.

Mutations in isocitrate dehydrogenase 1 or 2 (IDH1 and IDH2) are important to identify at the time of diagnosis of high- or very high-risk MDS. These particular mutations lead to abnormal leukemogenesis. Mutated IDH1 or IDH2 are not common and are only found in approximately 4–12% of patients with MDS. Those gene mutations have treatment impact. Recently, two IDH inhibitors, specifically ivosidenib targeting IDH1 and enasidenib for IDH2, are approved by the United States Food and Drug Administration (FDA) for use in AML, but not in MDS [27, 28].

Both agents are undergoing investigation in combination with azacitidine or with induction chemotherapy in patients with IDH-mutant MDS.

Other mutations are very important to identify early because of their prognostic impact, like SF3B1 mutations, in fact mutations of SF3B1 are strongly associated with ring sederoblasts, and a typical SF3B1 can be presumptive evidence of MDS, and have more favorable prognosis [29].

More than third of MDS patients with less than 5% of blasts will have an adverse gene mutation. These include mutations cited before like SRSF2, U2AF1, ASXL1, RUNX1, EZH1, TP53, IDH1, NRAS, and PRPF8, but the only mutation having good prognosis is SF3B1 mutation [30].

For patient with MDS and having more than 5% of blasts (5–30% blasts), several mutated genes retain their in higher risk MDS. In fact, mutation of TP53, CBL, RUNX1, PRPF8 are utch more common and remain adverse, and SF3B1 mutation are rare end no longer favourable.

Somatic mutations alone are not great predictors of outcomes after treatment with approved MDS therapies, but mutations of TP53 and epigenic regulators like TET2 and DNMT3A have shown associations with response to hypomethylating drugs in some studies. In contrast of that, we do have a cytogenical marker there is very good for predictive response to therapy: it s about Del (5q) and lenalidomide. In fact patient having Del(5q) can response favourably to lenalidomide, if TP53 mutations are absents, because TP53 mutations indicate resistance to lenalidomide and predict relapse or progression even after allogeneic stem cell transplantation.

Data are accumulating to support use of next-generation sequencing (NGS) in the diagnosis and management of patients with MDS.

The treatment and management of older patients with MDS is extremely challenging due to a number of reasons, including advanced disease, intolerability to therapy, significant comorbidities, and potential for more drug–drug interactions with concomitant therapy.

7. Conclusion

Our knowledge about the genetics of myelodysplastic syndromes (MDS) and related myeloid disorders has been dramatically improved during the past decade, in which revolutionized sequencing technologies have played a major role.

Cytogenetic abnormalities have extensive utility in MDS, they have many implications for diagnosis and prognosis. The best example is represented by MDS with isolated del(5q). the presence of del(5q) is known to be a lenalidomide-responsive condition with a clearly elucidated molecular mechanism.

The use of additional genomic information, provided by DNA microarrays and sequencing, holds great promise in further refining the classification and management of these disorders.

At present, NGS is rarely incorporated into clinical guidelines although an increasing number of studies have demonstrated the benefit of using NGS in the clinical management of MDS patients [31].

Conflict of interest

Authors declare have no conflict of interest.

Author details

Mounia Bendari[1*] and Nisrine Khoubila[2]

1 Mohammed VI University of Health Sciences, Casablanca, Morocco

2 Faculty of Medicine and Pharmacy of Casablanca, Hassan II University, Morocco

*Address all correspondence to: bendarimounia@gmail.com

References

[1] Tefferi A, Vardiman JW. Myelodysplastic syndromes. N Engl J Med. 2009; 361:1872-85.

[2] Nimer SD. Myelodysplastic syndromes. Blood. 2008;111:4841-51. 16. Ogawa S. Genetics of MDS. Blood. 2019;133:1049-59.

[3] Mufti GJ Pathobiology, classification, and diagnosis of myelodysplastic syndrome Best Pract Res Clin Haematol 2004 Dec;17(4):543-57

[4] Genovese G, Kahler AK, Handsaker RE, et al. ¨ Clonal hematopoiesis and blood-cancer risk inferred from blood DNA sequence. N Engl J Med. 2014;371(26):2477-2487.

[5] Vogelstein B, Papadopoulos N, Velculescu VE, Zhou S, Diaz LA Jr, Kinzler KW. Cancer genome landscapes. Science. 2013; 339(6127):1546-1558.

[6] Ades, L.; Itzykson, R.; Fenaux, P. Myelodysplastic syndromes. Lancet 2014, 383, 2239-2252.

[7] Greenberg PL, Tuechler H, Schanz J et al. Revised international prognostic scoring system for myelodysplastic syndromes. Blood 2012; 120: 2454-65.

[8] Greenberg, P.; Cox, C.; LeBeau, M.M.; Fenaux, P.; Morel, P.; Sanz, G.; Sanz, M.; Vallespi, T.; Hamblin, T.;Oscier, D.; et al. International scoring system for evaluating prognosis in myelodysplastic syndromes. Blood 1997, 89, 2079-2088

[9] Greenberg, P.L.; Tuechler, H.; Schanz, J.; Sanz, G.; Garcia-Manero, G.; Sole, F.; Bennett, J.M.; Bowen, D.; Fenaux, P.; Dreyfus, F.; et al. Revised international prognostic scoring system for myelodysplastic syndromes. Blood 2012, 120, 2454-2465.

[10] McQuilten ZK, Sundararajan V, Andrianopoulos N et al (2015) Monosomal karyotype predicts inferior survival independently of a complex karyotype in patients with myelodysplastic syndromes. Cancer 121:2892– 2899.

[11] Solé F, Luño E, Sanzo C et al (2005) Identifi cation of novel cytogenetic markers with prognostic signifi cance in a series of 968 patients with primary myelodysplastic syndromes. Haematologica 90:1168-1178

[12] Smith SM, Le Beau MM, Huo D et al (2003) Clinical-cytogenetic associations in 306 patients with therapy-related myelodysplasia and myeloid leukemia: The University of Chicago series. Blood 102:43-52.

[13] T. Comont et al, Management of myelodysplastic syndromes in 2019: An update La Revue de médecine interne 40 (2019) 581-589

[14] Fenaux P, Mufti GJ, Hellstrom-Lindberg E, Santini V, Finelli C Efficacy of azacitidine compared with that of conventional care regimens in the treatment of higher-risk myelodysplastic syndromes: a randomised, open-label, phase III study

[15] Bennett JM, Byrd J, Backstrom J, Zimmerman L, McKenzie D, Beach C, Silverman LR; International Vidaza High-Risk MDS Survival Study Group Lancet Oncol 2009 Mar;10(3):223-32

[16] Mario Cazzola, Matteo G. Della Porta, Luca Malcovati The genetic basis of myelodysplasia and its clinical relevance Blood. 2013 Dec 12; 122(25): 4021-4034. Prepublished online 2013 Oct 17. doi: 10.1182/ blood-2013-09-381665

[17] Yoshida K, Sanada M, Shiraishi Y, Nowak D, Nagata Y, Yamamoto R, et al. Frequent pathway mutations of

splicingmachinery in myelodysplasia. Nature. 2011; 478:64-9.

[18] Haferlach T, Nagata Y, Grossmann V, Okuno Y, Bacher U, NagaeG, et al. Landscape of genetic lesions in 944 patients with myelodysplastic syndromes. Leukemia. 2014;28:241-7.

[19] Graubert TA, Shen D, Ding L, Okeyo-Owuor T, Lunn CL, Shao J, et al. Recurrent mutations in the U2AF1 splicing factor in myelodysplastic syndromes. Nat Genet. 2011;44:53-7.

[20] Makishima H, Yoshizato T, Yoshida K, Sekeres MA, RadivoyevitchT, Suzuki H, et al. Dynamics of clonal evolution in myelodysplastic syndromes. Nat Genet. 2017;49:204-12.

[21] Delhommeau F, Dupont S, Valle VD, James C, Trannoy S, Massé A, et al. Mutation in TET2 in myeloid cancers. New Engl J Med. 2009; 360:2289-301.

[22] Mardis ER, Ding L, Dooling DJ, Larson DE, McLellan MD, Chen K, et al. Recurring mutations found by sequencing an acute myeloid leukemia genome. N Engl J Med. 2009; 361:1058-66.

[23] Nikoloski G, Langemeijer SM, Kuiper RP, Knops R, Massop M,Tönnissen ER, et al. Somatic mutations of the histone methyltransferase gene EZH2 in myelodysplastic syndromes. Nat Genet. 2010; 42:665-7.

[24] 14. Lee, E.J.; Podoltsev, N.; Gore, S.D.; Zeidan, A.M. The evolving field of prognostication and risk stratificationin MDS: Recent developments and future directions. Blood Rev. 2016

[25] Meggendorfer, M.; de Albuquerque, A.; Nadarajah, N.; Alpermann, T.; Kern,W.; Steuer, K.; Perglerova, K.;Haferlach, C.; Schnittger, S.; Haferlach, T. Karyotype evolution and acquisition of FLT3 or RAS pathwayalterations drive progression of myelodysplastic syndrome to acute myeloid leukemia. Haematologica2015, 100, e487–e490.

[26] Gill, H.; Man, C.H.; Ip, A.H.; Choi,W.W.; Chow, H.C.; Kwong, Y.L.; Leung, A.Y. Azacitidine as post-remission consolidation for sorafenib-induced remission of FMS-like tyrosine kinase-3 internal tandem duplication positive acute myeloid leukemia. Haematologica 2015, 100, e250–e253.

[27] Gil-Perez A, Montalban-Bravo G. Management of myelodysplastic syndromes after failure or response to hypomethylating agents. Ther Adv Hematol. 2019; 10:2040620719847059.

[28] DiNardo CD, Jabbour E, Ravandi F, et al. IDH1 and IDH2 mutations in myelodysplastic syndromes and role in disease progression. Leukemia. 2016;30(4):980-984.

[29] Arber A. The 2016 revision to the World Health Organization classification of myeloid neoplasms and acute leukemia *Blood* (*2016*) *127* (*20*): *2391*-2405

[30] Bejar R. aSomatic Mutations in MDS Patients Are Associated with Clinical Features and Predict Prognosis Independent of the IPSS-R: Analysis of Combined Datasets from the International Working Group for Prognosis in MDS-Molecular Committee *Blood* (2015) 126 (23): 907

[31] Tobiasson M A. O. Kittang3 Treatment of myelodysplastic syndrome in the era of next-generation sequencing M. 2019 The Association for the Publication of the Journal of Internal Medicine

Chapter 10

Research of the Philadelphia Chromosome in Chronic Myeloid Leukemia: Diagnostic and Prognostic Interests

Yahya Benbouchta, Ahmed Afailal Tribak and Khalid Sadki

Abstract

Myeloproliferative syndromes are cell proliferation involving one or more medullary lines without blocking maturation. Chronic myeloid leukemia (CML) is the most common of these syndromes, it corresponds to the monoclonal proliferation of a multipotent stem cell; the myeloblastic or lymphoblastic transformation of CM. has a poor prognosis. The Philadelphia chromosome t(9;22)(q34;q11) is the first cytogenetic abnormality that has been associated with a malignant process. It is found in 89 to 95% of CML. The search for the Philadelphia chromosome (Ph1) has multiple interests: Diagnostic, prognostic and in therapeutic monitoring. The search for the Philadelphia chromosome by molecular cytogenetics makes it possible to remedy the poverty of cell suspensions in metaphase to take up the inconclusive results of classic cytogenetics on nuclei in interphase and to detect residual disease during therapeutic monitoring. Through the literature review, we highlight the importance of the identification of the Philadelphia chromosome in Myeloproliferative Syndromes for the improvement of the quality healthcare of the affected patients.

Keywords: Philadelphia chromosome, myeloproliferative syndromes, karyotype, diagnostic and prognostic interest

1. Introduction

Leukemias are clonal and acquired diseases of the hematopoietic stem cell or a precursor already committed to lymphoid and /or myeloid lineages [1]. hyperplasia produced a tissue that results from cell proliferation as myeloid pathology. Chronic myeloid leukemia (CML) is a monoclonal pathology of the pluripotent stem cell characterized by neoplastic granulocytic overproduction. This myeloproliferative syndrome has two particular characteristics:

- Its evolutionary mode consists of a chronic chemosensitive phase, followed by an acceleration phase, then an acute (or blast) transformation, ineluctable and chemoresistant.

- A quasi-constant clonal cytogenetic marker which is the Philadelphia Ph1 chromosome or derived from chromosome 22. This chromosome abnormality is generated from the reciprocal translocation involving the q34 band of chromosome 9 and the q11 band of chromosome 22.

The recent development of therapeutics targeted at the activity or stability of an oncogenic protein has recently been illustrated by the therapeutic successes obtained in the treatment of chronic myeloid leukemia and acute promyelocytic leukemia [1]. Until now cytogenetics has been the reference for structural abnormalities, in particular translocations, tools for precise diagnosis in certain disputed cases and the detection of residual diseases or possible relapses. However molecular cytogenetics can detect chromosomal abnormalities of small sizes not visible on metaphasic chromosomes (semi-cryptic). It is of particular interest in the analysis of acquired abnormalities and is involved in monitoring the persistence of an abnormal clone in order to detect predicted recurrent translocations and may also help characterize genes in the evolutionary process of carcinogenesis. The current recommendations are based on high-quality evidence reported in peer-reviewed journals, supplemented by expert group consensus. These recommendations apply to healthcare professionals who treat CML patients and CML patients to better understand their conditions and treatments [2].

2. Interest of chromosome Philadelphia in chronic myeloid leukemia

- The usual form or standard translocation.

It is the translocation of a distal fragment of the long arm of chromosome 22 (fragment 22q11.2) to the distal part of the long arm of chromosome 9 with recovery of a deleted part of the long arm of chromosome 9 on the long arm of chromosome 22. It is therefore a reciprocal translocation, without loss of chromosomal material (**Figure 1(a,b)**).

Since this date, we defined the standard Philadelphia chromosome as: **t(9;22)(q34;q11) or t(9; 22)(q34.1; q11.21)**.

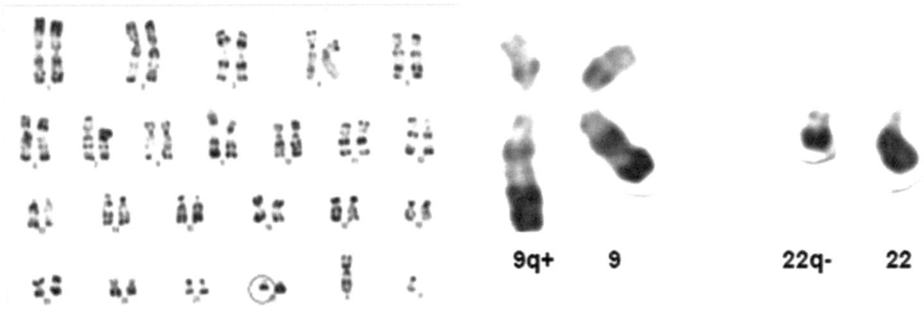

Figure 1.
(a) Result of a metaphase karyotype not classified in the R-band. (the circle indicates the Ph1 chromosome).
(b) Partial RHG band karyotype of one of our patients: t(9;22)(q34;q11).

3. Diagnostic interest of chromosome Philadelphia in chronic myeloid leukemia

3.1 The chronic phase

The Philadelphia chromosome is the only element allowing a diagnosis in hyperleukocytosis. It is found in 89 to 95% of CML cells: In the granulomonocytic, erythroblastic and B lymphocytic lines [3]. In most cases, CML is diagnosed on clinical and hematologic data alone. The differential diagnosis arises with all the pathologies that are accompanied by hyperleukocytosis with mild myelemia.

- The Ph1 chromosome: Diagnoctic key

The almost constant presence of this translocation in CML offers clinicians an additional diagnostic tool especially in myeloproliferative syndromes (MPS), chromosome 22 can be translocated to a chromosome other than chromosome 9 or else participates in a complex translocation of most interest, often three chromosomes of which the 22 and 9 one speaks then of Ph1 variant as opposed to the standard Ph1 chromosome. This same tanslocation t(9;22)(q34;q11) is found in a non-negligible percentage in ALL and AML.

3.2 Differential diagnosis in the acute phase

In acute leukemia, there is an accumulation of immature precursors of the hematopoietic lineage involved in the Bonne marrow, blood, or other tissue pathologies. The acutization phase CML disease there is a significant hyperleukocytosis with the presence of the Philadelphia chromosome on all mitoses. This acutization phase is preceded by the appearance of secondary anomalies: Trisomy 8, duplication of Ph1, and isochromosome 17, which conditions a poor prognosis.

We also find the Philadelphia chromosome:

- In 5% of acute lymphoblastic leukemia (ALL) in children and 20–30% of ALL in adults and also found in acute myeloid leukemia, In acute myeloid leukemia type 1 (LAM1) and LAL1 [4].

3.3 Other myeloproliferative syndromes

Essential thrombocythemia, myeloid splenomegaly, polycythemia vera or vaquez disease and chronic myelomonocyte leukemia (CMML) have the same phenotype as show in certain forms of CM. For this reason, it is important to confirm the diagnosis of chronic myelogenous leukemia by cytogenetic study or molecular biology [5]. Sometimes to give a right diagnosis is complicated so only the karyotype or molecular biology tests can help for that. The first test looks for the presence or not of the Ph1 while the other molecular biology tests investigate the *BCR-ABL* rearrangement.

3.4 Chronic myeloid leukemia in children

Chronic myeloid leukemia in children: There are two clinically and genetically distinct forms:

- The adult form occurring beyond the age of two years resembles in all respects a Ph1 + CML with the presence of the cytogenetic marker Ph1 + and break points in M-*BCR* especially in 5' [6, 7].

- The juvenile form before the age of two characterized by a peculiar clinical picture and a normal karyotype in most cases otherwise the most frequent chromosomal aberration is monosomy 7.

In some cases, the Ph1 chromosome may be masked due to the size of the fragment translocated which is submicroscopic, molecular cytogenetics are then used

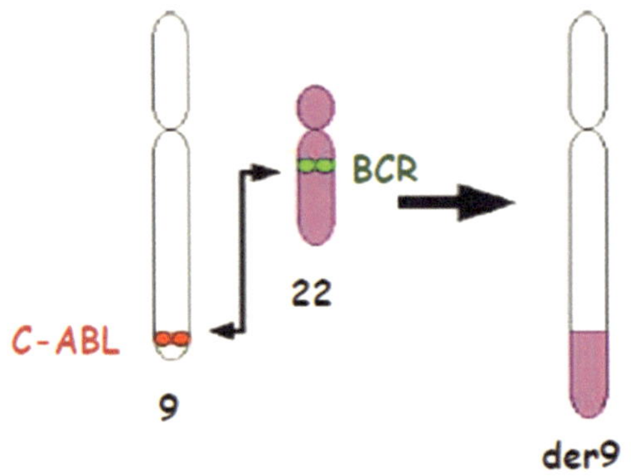

Figure 2.
The Philadelphia chromosome can be masked because of the size of the translocated fragment which is submicroscopic.

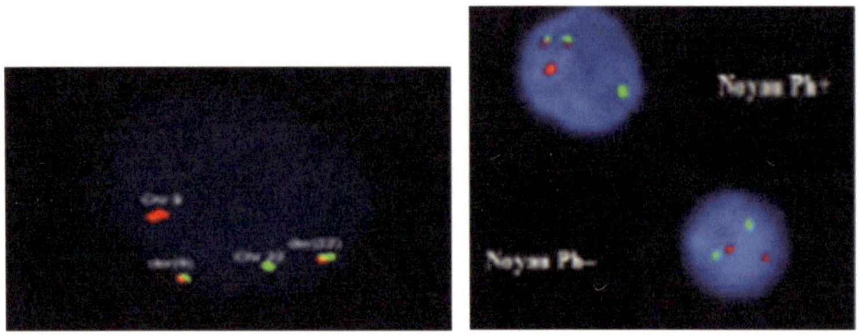

Figure 3.
FISH image of BCR/ABL *positive rearranged metaphase.*

(in situ hybridization: FISH) or real-time PCR search for the Philadelphia chromosome is necessary to confirm the diagnosis of CML and to monitor progress under certain anti-mitotic drugs (**Figure 2**).

In onco-hematology, FISH provides a decisive complement to the diagnosis, the prognosis and monitoring of targeted therapies. In leukemia chronic myeloid this technique highlights the fusion of genes *BCR* and *ABL* which characterize the Philadelphia chromosome (Ph). FISH is particularly interesting in the cytogenetic monitoring of CML. In due to culture problems (low mitotic index and the quality of the metaphases poor according to European Leukemia Net 2009. This service is currently offered to patients with CML as part of the cytogenetic monitoring of their disease (**Figure 3**).

3.5 Variant translocations

Variant translocations fall into two subgroups: Simple variants and complex variants; their definitions are based on the results of R, G banding and molecular biology. Although it is very common, it is quickly learned that the t(9;22) translocation is not pathognomonic for CML and it has several variants: the Ph1 (+) variants, the masked Ph1 chromosome and the Ph1 (−) variants. All chromosomes except Y are involved in the variant form of Ph1 especially chromosomes 3, 11, 12, 14 and 17 [8]. The variants can all be considered as complex translocations since the molecular genetic investigations of the supposed simple variants show that they involve at least three chromosomes and always the 9 and the 22 [9].

3.6 The blast transformation

In this phase, 65 to 80% of patients develop additional chromosomal aberrations not due to chance which precede clinical and hematological manifestations by several months and which can serve as indicators prognosis [10, 11]. Secondary anomalies appear: Double chromosome Philadelphia, trisomy 8, isochromosome 17 and trisomy 19. These four additional abnormalities are part of the clonal course in 70% of CML Other, more rarely encountered anomalies seem to be due to chance, thus taking the minor pathways. In more than 50% of cases, they are represented by:

- Monosomies: Y, 7, 17.
- Down's syndrome: 17 and 21.
- And the translocation t(3;21)(q26;q22) which has the characteristic of being accompanied by medullary fibrosis [12].

A quarter of patients [10, 11] will not develop any additional abnormalities and will keep Philadelphia alone for the duration of their survival.

3.7 Chronic myeloid leukemia with secondary abnormalities

The following partial karyotypes show the association of certain additional abnormalities to the Philadelphia chromosome (Ph1) in our patients. However, the therapeutic and prognostic approach is totally different. It is therefore necessary:
Make a positive diagnosis for CML.

- Correct the diagnosis of certain contentious cases.
- Specify the evolutionary stage.
- And make a differential diagnosis with myeloprolifirative and myelodysplastic syndromes.

During the blast phase of CML at Ph1 (+), analysis determines as a factor of poor prognosis [10]. As for the Philadelphia chromosome alone, it appears to have an independent prognostic value [13].

4. Prognostic interest of chromosome Philadelphia in CML

Evaluating the prognosis of CML using clinical-biological criteria can predict the probable date of onset of blast transformation which amounts to determining the probable duration of the chronic phase. As regards the cytogenetic criterion, it must be defined and homogeneous. The prognoses of Ph1 (+) CML and Ph1 (−) CML should be studied separately because we have seen the current difficulties of including the Ph1 (−) form in the nosological framework of CML.

In our medical genetics' laboratory. The suspected diagnosis was CML in 69 patients, unlabeled SMP in the remaining 22 patients (**Table 1**).

- Culture failure in 6 cases.
- Normal karyotype in 25 cases.
- Philadelphia chromosome or t(9;22)(q34;q11) in 60 cases.

The cytogenetic criterion is requested at two levels:

- For the initial assessment of prognosis at the time of diagnosis of CML in combination with baseline clinical and hematologic data.
- Then to assess the prognosis later during the blast transformation.

During the chronic phase of Ph1 (+) CML and without the knowledge of multi-parametric analyzes, it has been shown that the most significant prognostic factors which determine the duration of survival are [14]:

- The presence of additional clonal chromosomal abnormalities (relative risk "RR" = 4.5).
- Circulating blasts greater than 5% "RR = 1.8".
- A hemoglobin rate of less than 10 g / dl "RR = 1.30".

Results / Indications	Ph(+)	Ph(-)	Culture failure	Total
Chronic myeloid leukemia	54	12	03	69
Myeloprolofrative syndrome	06	13	03	22

(a)

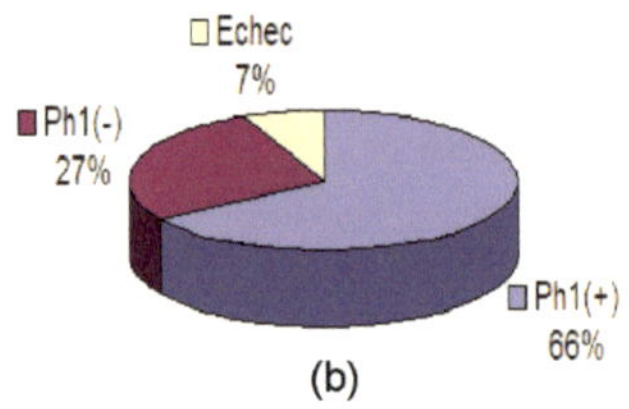

(b)

Table 1.
(a) Cytogenetic analysis in myeloproliferative syndromes 91 patients. (b) Frequency of the Philadelphia chromosome (Ph1).

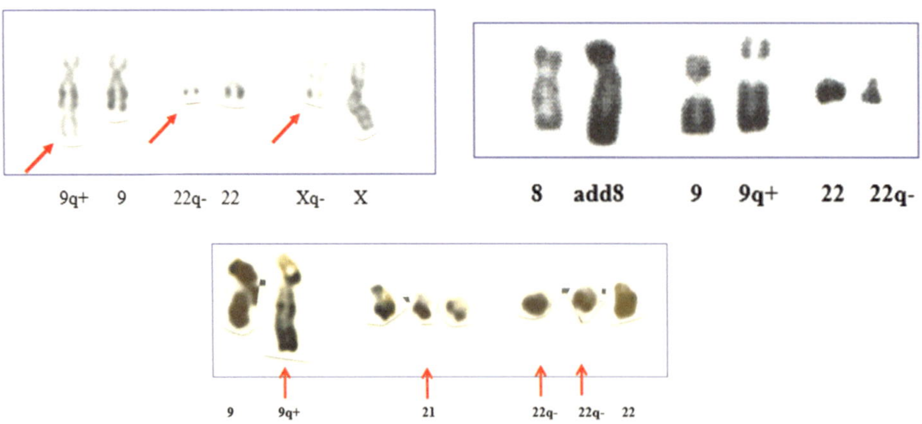

Figure 4.
(a) Presence of an extra Ph1 chromosome and trisomy 21. (b) Partial trisomy 8. (c) Partial band karyotype RHG: T(9;X;22).

- Thrombocytopenia.
- Leukocytosis>20.109 / l.
- Non-lymphoid blast cells.
- A clonal evolution a \ double chromosome Ph1, a trisomy 8, and typical aberrations of the acute phase (Ph1 (+), i(17q), hypodiploidy or hyperdiploidy).
- Lack of response to treatment.

It is interesting to note that the double chromosome Ph1 or trisomy 8 are more frequent in acute transformations of the AML, ALL type and that they respond poorly to treatment (**Figure 4**).

5. Surveillance of residual disease

Residual disease is defined as the number of malignant cells persisting after cytotoxic treatment, the eradicative action of which is intended to be as complete as possible: Chemotherapy, ionizing radiation, bone grafting. The residual malignant cells which escape this treatment can be the cause of a relapse hence the need to quantify them as precisely as possible. Before the introduction of molecular biology, hematologists had at their disposal various means of approach to define the biological remission of a hemopathy: Cytology, cytogenetics, immunology. In the best case, the sensitivity of these techniques did not make it possible to detect less than one residual cell in 100, a very insufficient sensitivity threshold to help clinician to decide for adequate treatment and to evaluate the quality control of the graft. The treatments envisaged must be carried out in order to obtain hematological remission and if possible a complete eradication of the Ph1 (+) cells (cytogenetic remission) with regard to chronic myeloid leukemia, the evolution takes place in two stages: A first chronic or myelocytic phase easily controlled by usual therapies then a second inconstant transition phase called acceleration with resistance to conventional chemotherapy, following which an acute transformation occurs, often of the terminal acute myelogenous leukemia type, constantly fatal, inevitable on average 3 to 4 years after diagnosis.

6. Contribution of oncocytogenetics in chronic myeloid leukemia

The molecular consequence is the formation of a *BCR-ABL* fusion gene, transcribed into 8.5 Kb mRNA and translated into 210 Kd protein with greater tyrosine kinase activity compared to the normal protein of the proto oncogene c-*ABL* from 145 Kb, this protein is involved in the pathological process of CML [15]. The molecular biology techniques applied to DNA, mRNA (RT-PCR) and encoded

Chromosomes	RNAm	Protein
9	ARNm c-abl	$P145^{c\text{-}abl}$
22	ARNm bcr	$P160^{bcr}$
22q-	ARNm bcr-abl	$P210^{bcr\text{-}abl}$

Table 2.
The chromosomes involved and their molecular consequences.

Figure 5.
Schema the Philadelphia chromosome t(9;22)(q34;q11) results in the fusion of BCR *genes on chromosome 22 and* ABL *on chromosome 22. The fusion protein has a strong activity tyrosine kinase responsible for tumor development.*

proteins have made it possible to specify the nature of the molecular events resulting from the rearrangement of *BCR-ABL* (**Table 2**).

Fluorescent in situ hybridization (FISH) using specific probes provides a useful tool for the detection of t(9;22)(q34;q11) and *BCR-ABL* rearrangement [16] (**Figure 3**). The fusion protein has a strong activity tyrosine kinase responsible for tumor development (**Figure 5**).

7. Conclusion

In summary, Philadelphia chromosome is an abnormal chromosome 22, resulting from a reciprocal translocation between chromosomes 9 and 22, a specific marker in chronic myeloid leukemia. His research in myeloproliferative syndromes has multiple interests: Diagnostic, prognostic and therapeutic follow-up which contributes to better patient care. Its demonstration in myeloproliferative syndromes makes it possible to confirm the nature of the disease and to distinguish between CM and other myeloproliferative syndromes.

Acknowledgements

The authors would like to thank the patients and their family. We are grateful to all of the staff of the Department of Medical Genetics of the National Institute of Health for their continuous support.

List of abbreviations

CML	chronic myeloid leukemia CML
Ph	Philadelphia
ALL	acute lymphoblastic leukemia
LAM	acute myeloid leukemia type 1
LAL1	acute lymphoblastic leukemia type 1
ALL	acute lymphoblastic leukemia
AML	acute myeloid leukemia
CMML	chronic myelomonocyte leukemia
MPS	myeloproliferative syndromes

Author details

Yahya Benbouchta[1,2*], Ahmed Afailal Tribak[2] and Khalid Sadki[2]

1 Department of Medical Genetics, National Institute of Health, Rabat, Morocco

2 Laboratory of Human Pathology, Faculty of Sciences, Mohammed V University in Rabat, Morocco

*Address all correspondence to: benbouchtayahya@yahoo.fr

References

[1] Gisselbrecht, S., *[Oncogenes and leukemia: history and perspectives]*. Med Sci (Paris), 2003. **19**(2): p. 201-210.

[2] Hochhaus, A., et al., *European LeukemiaNet 2020 recommendations for treating chronic myeloid leukemia.* Leukemia, 2020. **34**(4): p. 966-984.

[3] Shepherd, P.C., T.S. Ganesan, and D.A. Galton, *Haematological classification of the chronic myeloid leukaemias.* Baillieres Clin Haematol, 1987. **1**(4): p. 887-906.

[4] Médicale, G., *formelle, chromosomique, moléculaire, clinique. Par le collège National des Enseignants et Praticiens de Génétique Médicale. Coordonné par: Marc Jeanpiérre, Phlippe Jonveaux, Didier Lacombe, Nathalie Leporrier, Stanislas Lyonnet, Claude Mauraine*. 2004, Masson, Paris.

[5] Mahon, F.X. and J. Reiffers, *[Chronic myeloid leukemia. Diagnosis, course, prognosis]*. Rev Prat, 1996. **46**(18): p. 2231-4.

[6] Arlin, Z.A., R.T. Silver, and J.M. Bennett, *Blastic phase of chronic myeloid leukemia (blCML): a proposal for standardization of diagnostic and response criteria.* Leukemia, 1990. **4**(11): p. 755-7.

[7] Fioretos, T., et al., *Molecular analysis of Philadelphia-positive childhood chronic myeloid leukemia.* Leukemia, 1992. **6**(7): p. 723-5.

[8] Fitzgerald, P.H. and C.M. Morris, *Complex chromosomal translocations in the Philadelphia chromosome leukemias. Serial translocations or a concerted genomic rearrangement?* Cancer Genet Cytogenet, 1991. **57**(2): p. 143-51.

[9] Dobrovic, A., et al., *Molecular diagnosis of Philadelphia negative CML using the polymerase chain reaction and DNA analysis: clinical features and course of M-bcr negative and M-bcr positive CML.* Leukemia, 1991. **5**(3): p. 187-90.

[10] Griesshammer, M., et al., *Chronic myelogenous leukemia in blast crisis: retrospective analysis of prognostic factors in 90 patients.* Ann Hematol, 1996. **73**(5): p. 225-30.

[11] Kantarjian, H.M., et al., *Chronic myelogenous leukemia in blast crisis. Analysis of 242 patients.* Am J Med, 1987. **83**(3): p. 445-54.

[12] Heim, S. and F. Mitelman, *Cancer cytogenetics: chromosomal and molecular genetic aberrations of tumor cells*. 2015: John Wiley & Sons.

[13] Tanaka, K., et al., *Influence of M-BCR breakpoint sites on the duration of chronic phase in 100 patients with chronic myelocytic leukemia.* Cancer Genet Cytogenet, 1993. **70**(1): p. 39-47.

[14] Ariad, S., et al., *Prognostic factors in chronic myeloid leukaemia--importance of staging or disease biology.* S Afr Med J, 1992. **81**(6): p. 299-303.

[15] Rowley, J.D., *Chromosome abnormalities in leukemia and lymphoma.* Ann Clin Lab Sci, 1983. **13**(2): p. 87-94.

[16] Martinet, D., D. Muhlematter, and M. Jotterand Bellomo, *[Fluorescent in-situ hybridization technique (FISH) in the diagnosis of Philadelphia translocation in chronic myeloid leukemia]*. Schweiz Med Wochenschr, 1996. **126**(20): p. 855-63.

Chapter 11

Contribution of Environmental Constituents in the Genomic Disruption of Cytokeratins

Vishnu Sharma, Tarun Kr. Kumawat, Garima Sharma, Rashi Garg and Manish Biyani

Abstract

Cytokeratins are keratinous protein and assist cells to reduce mechanical stress on the intracytoplasmic layer of epithelial tissue. There are several unspecified mutations in the epithelial layer that may induces by environmental mutagens and pathogens. The unspecified mutations in the epithelium surface also disrupt biology of skin at multiple different levels and cause innate keratinizing disorders. These serve as a root generator of neurohormones and neuropeptides which mainly partake in the disruption. Generally, all 54 unique genes of human keratin partake in mutations and cause cutaneous tissue fragility, skin hypertrophic, and malignant transformation. In this chapter, unspecific factors that involved in the pathogenesis of skin diseases and the ways by which such keratin changes might harness to alleviate different skin conditions are also included. Consequently, the contribution of environmental changes in the frontier of mutations or misregulations of the cytokeratin genes, is also cited here.

Keywords: keratins, skin barrier, keratinopathies, keratinocytes, protease allergens

1. Introduction

Cytokeratin is an intermediate keratinous protein found in the intracytoplasmic cytoskeleton of epithelial tissue. Cytokeratin possess essential components for the epithelium layer with diameters of 6 nm (microfilaments) to 25 nm (microtubules). The microfilaments or microtubules support cells to resist mechanical stress [1–3]. In the 1970s, the term 'cytokeratin' was derived through the protein characterization in the intermediate filament [4]. Later, in 2006, with the new systemic nomenclature, the terminology "cytokeratin" was termed as keratins. This nomenclature was under the nomenclature of the Human Genome Organization (HUGO) for both the gene and protein names [5, 6].

Keratin are the most complex protein in vertebrates and in filamentous form are essential epithelial cell structural stabilizers. That's why; keratin is of unprecedented importance in genetics, embryology, pathology, and dermatology [7]. Keratin filaments are usually integrated with desmosomes and hemidesmosomes. They contribute: to the cohesion of the epithelial cells; for attachment of the basement membrane; and for epithelial connectivity tissue transition [8, 9]. Although few keratin structures/skeletons also found as scattered or dispersed between

 IntechOpen

keratinous filaments in the cytoplasm of the internal parenchymatous organs. This behavior contributes to the (simple) unstratified epithelial membrane. These component sprouts increase as tonofilaments and transformed into a cornified stratified epithelial sheet [10].

Although, Keratin are classified into alpha and beta keratins according to the amount of sulfur content and structure [11]. Alpha (α) keratin form epithelial layers in all vertebrates [12, 13]. Configurationally, alpha (α) keratin has abundant amounts of the hydrophobic amino acids: methionine, phenylalanine, valine, isoleucine, and alanine. Due to the occurrence of these amino acids, alpha (α) keratin is extraordinary for its strength, elasticity, tiredness, insolubility, and durability [14]. Apart from the diversity of the epithelial keratins ('soft' or 'cyto'); hair and nails are built from a very distinct subfamily of 'hard' or 'trichocytic keratins' [10, 15–18]. Since they are enriched in stacked β pleated sheets and are known as corneous beta-protein" or "keratin-associated beta-protein [19–20]. The epithelial keratins differ in their non-α-helical head and tail domain. This is due to the existence of high sulfur contents which is primarily responsible for the high filament linkage level of the keratin-associated protein [21, 22].

In filamentous form, keratin possesses a head-rod-tail structural arrangement with its basic polypeptide configuration consisting of a core alpha-helical coiled rod structure of about 310 amino acids in size. This central rod segment contains four helical structures interrupted by three short non-helical flexible reticulation/Linker regions [9, 12, 23]. These linker or reticulation regions flank by the complex, non-helical amino-terminal head and carboxy-terminal tail domains. In the rod domain, a heptad repeat of amino acid residues is present. Furthermore, near the middle of the domain, the "stutter" region is found and is a highly conserved segment among IFs. This area does not take part in the development of coiled-coil heterodimers but plays specific roles in the extension and rotation characteristics of keratins [24, 25]. At the beginning and end of the heptad repeat regions in the rod domain; are the highly conserved helix initiation motif (HIP) and helix termination peptides (HTP). Respectively, HIP & HTP, both consist of 20 amino acid sequences related to different keratin gene families [26–28]. The heteropolymeric structure of type I and type II keratin collectively generate a filamentous form. These are all aligned laterally with a scalable and parallel overlap and form KIFs through lateral stacking and polymerization of chains [9, 29, 30].

In cytokeratins, all of the intermediate filament proteins have the template/prototype structure that contains a central coiled-coil helix (310–350 amino acids). These are surrounded by the variable-length globular NH2-terminal head and COOH end-end tail domains [31–33]. Here, the keratin type I chain form a paired dimer with its type II counterpart and build an antiparallel fashion to a tetramer. Two tetramers transversely bind and resulted in a protofilament. Then, protofilaments are twisted into a keratin filament rope. Therefore, each keratin filament has a cross-section of 32 individual helical coils (**Figure 1**) [34–38]. The globular end-domains in most of the intermediate filament proteins contain all known sites for phosphorylation, glycosylation, and other relative activities [33, 39–42].

Usually, cytokeratin has an outstanding standard due to its high molecular diversity. The molecular weight of human keratins ranges from 44 to 66 kDa. In human, all 54 distinct functional keratin genes are found on chromosomes 12 & 17 and represent the typical intermediate filament category of epithelial cells. All keratin filaments are specifically bundled as tonofilaments in some but not all endothelium [43, 44]. In humans, all 54 distinct functional keratin genes are characterized as keratin type- I genes (17 epithelial and 11 hair keratin genes) and keratin type- II genes (20 epithelial keratins and 6 hair keratins) (**Figure 2**) [45, 46]. Type I

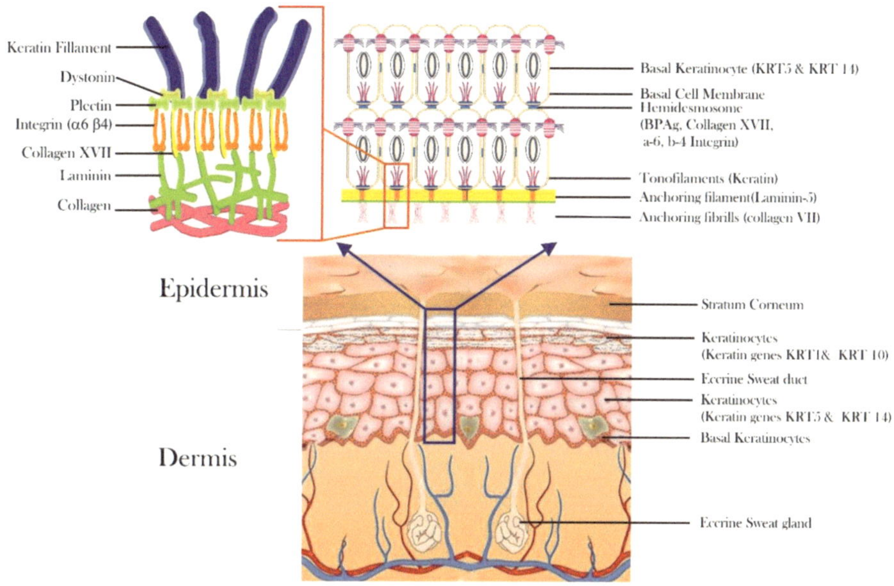

Figure 1.
Differentiated Keratinocytes in Structure of Human Skin.

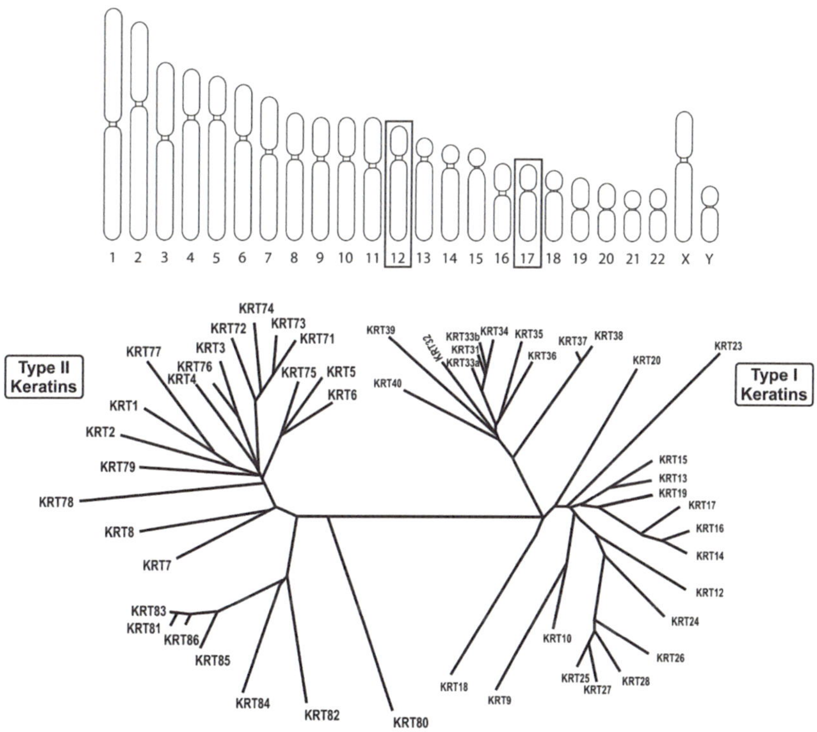

Figure 2.
Ideogram and Phylogenetic description of Mammalian Keratin Chromosome (i.e. 12 & 17).

keratin comprises the KRT9–KRT10, KRT12–KRT28, and KRT31–KRT40 (including KRT33a and KRT33b) genes, while type II keratin has KRT1–KRT8 (including KRT6a, KRT6b, and KRT6c) and KRT71–KRT86. Besides, the "stratified keratins" comprise KRT1 to KRT6 and KRT10 to KRT17 [10, 47–51].

The human skin is the largest of the epithelial layers of the body and has a crucial strategic defensive position at the boundary between the interior and exterior environment of the body. Histologically, the skin is made of multi-layered, non-vascular, stratified epidermal squamous tissues which are extending to dense fibrous connective dermal tissues [52, 53]. For each cytoskeleton of the epithelial layer, there are three abundant filament systems available: actin-microfilament systems (MFs; 7–10 nm diameter; IFs; 10–12 nm diameter), interlinked microtubules, and (MTs; diameter 25 nm). Each filament system comprises a corresponding gene family with specially regulated cell-tissue regulators that encode each protein family [54]. Mutations in any cutaneous-associated keratin genes are increasingly apparent, which causes a host of hereditary skin disorders; and specified to weakened cell tissue integrity & damaged the skin [42, 55–57].

The regional specificity of keratin expression may add to the intrinsic specialization of regional keratinocyte stem cells. Disorders in keratin may be genetic or acquired. Several keratin mutations have been identified as a cause for many diseases in the skin and mucosal tissues [26, 58–60]. Beyond the biological roles, keratin expression describes cells not only as "epithelial," but as distinctive even for characterize stages of cellular epithelial differentiation from embryonic to adult or internal maturation programs throughout growth. [10, 61, 62].

Mutations in the keratin amino acid sequence significantly affect the montage of keratin filaments [63–65]. Keratin genes repeat along with frameshift mutation and generate the pseudogenes. Approximately 87% of individuals' keratin pseudogenes are equivalent to keratin genes 8 & 18 [66]. keratin pseudogenes continuously exchange their position on multiple chromosomes through crossing-overs or translocations [45, 67]. Mutations of keratin genes contribute to several human and murine skin disorders [68].

Keratinous genetic mutations cause various keratin disorders known as "Keratinopathies". In the last decades, the spectrum of skin disorders has been increased enormously due to the abnormal function of structural proteins (keratins, filaggrin, loricrin, cornified cell envelope, etc.) [26, 47, 69–73]. It has also boost up or affected by deficient enzymes or transport proteins that are essential for lipid metabolism in the epidermis (cholesterol sulphatase, lipoxygenases, ABCA12, etc.). Mutations in epidermal keratins cause several skin diseases like congenital ichthyosis, epidermolytic ichthyosis, congenital bullous disease, corneal dystrophy, erythroderma, and pachyonychia congenital [74–76]. All the above diseases are due to mutation in the corneal keratin genes (KRT3, KRT12); mutation in simple epithelial keratin genes (KRT8/18, KRT19, KRT9); mutations in epithelial keratin genes (KRT6A, KRT6B, KRT6C, KRT16 or KRT17); mutations inKRT1 or KRT10 or basal layer keratinocytes genes (KRT5&KRT14) [77, 78].

In the epidermis and associated skin appendages, mutagenic cutaneous disorders are commonly termed as genodermatoses [79]. Keratin mutations represent keratin-related disorders including epidermolysis Bullosa Simplex (Keratin gene 5& 14), Keratinopathic Ichthyosis (Keratin gene l, 2 &10), Palmoplantar Keratoderma (Keratin gene 9), Pachyonychia Congenital (Keratin gene 6a,6b1,16 &17), and Monilethrix (Keratin gene 81, 83 & 86), etc. [18, 26, 80–82].

Epidermolysis bullosa is one of hereditary keratin disorder specified as mechanic-blistered skin. Epidermolysis bullosa is caused by desmoplakin or plakophilin type mutations in keratin genes KRT5 and KRT14 [83, 84]. In the manifestation of Epidermolysis bullosa, skin fragility reveals and can be increased in fluid-filled blister form or by the erosion of the skin. It causes failure in keratinization and affects the integrity or the ability of the skin to resist mechanical stresses. Hence, disease signs may extend for esophageal contraction, squamous carcinoma [85].

Epidermolysis Bullosa possess four main types: Epidermolysis Bullosa Simplex (EBS), Dystrophic Epidermolysis Bullosa (DEB), Junctional Epidermolysis Bullosa (JEB), and Kindler Syndrome (KS). In Epidermolysis Bullosa, all above disorders are inherited keratin disorder caused by mutations in Keratin genes 5 & 14 and plectin [86, 87]. The mutations in the gene of plectin are associated with muscular dystrophy. Symptomatically, Suprabasal epidermolysis causes fragility and blistering of skin or erosion by minor injury or friction (Rubbing or Scratching). Blistering may extend to mucous of the mouth with digestive tract and directly can affect the digestive system. Because of this, many infected children are poorly nourished and grow variably. The extended blistering cause irregular red patches of granulation sheath and rise to bleed regularly. In the enlargement of the disease, newborns lose essential metabolic nutriment and fluids. Consequently, granular sheath affects the respiratory tract and difficult to speaking and breathing [88, 89]. Besides, Kindler syndrome is a type of epidermolysis bullosa and causes skin blistering but often on the hands and feet. It generates scarring on the skin between the fingers and between the toes. Kindler syndrome is genetic dermatitis even though may also cause by ultraviolet (UV) rays and sunburn undoubtedly. The Kindler syndrome can expand on the outward of the oral cavity, throat, intestines, genitals, and urinary regions. Ultimately, the condition might be converted into squamous follicle melanoma [90–93].

Keratinopathic ichthyosis is a generative disorder in the human that occurs by a mutation in Keratin genes (KRT1, KRT2, and KRT10). It exhibits anomalies in the membranous filaments and develops the spectrum of clinical manifestations [26, 94, 95]. Therefore, it is also known as superficial keratin keratodermas [60, 96]. This disorder manifests symptoms as scratched skin fragility, blister generation over flexural areas on erythroderma, and thick stratum corneum. The blistering and erythema in Neonates appear by birth though over time manifestation increases. Usually, Palmoplantar keratoderma also appears with Keratinopathic ichthyosis. It exhibits the thickening of the palm's skin of the hands and soles of the feet [26, 97, 98].

Monilethrix is a pervasive innate disorder. It occurs due to mutation in the human keratin gene KRT86 and KRT81 [99–100]. Clinically, Monilethrix is distinguished by dystrophic hair reduced region or with complete alopecia. Hair shaft deformation is defined by elliptical nodes that are commonly separated by reductions in skin color, by the appearance of scars, scratches, or rashes at the constricted regions [101, 102]. Infected hair also exhibits scarcities in the cortex of the skeletal hair shaft proteins, especially for trichocyte keratins. Rarely, hair possesses regrowth through adolescence or pregnancy [103, 104].

Pseudofolliculitis Barbae (PFB) and loose anagen hair syndrome (LAHS) are other hair disorders due to mutation in the keratin gene KRT75. Certain gene is located in the form of a cluster on the long arm of chromosome twelve. It originates with the follicular infection on the neck and beard region of the face [105–108]. Sometimes disease also expands due to shaving on surrounding ingrown facial hairs, on the body wherever hair is shaved or plucked, including axilla, pubic area, and legs [18]. Apart from the cutaneous layer, nail dysplasias (Pachyonychia congenital) also take place in the hypertrophic portion of nails. It occurs due to mutation in the genes KRT6a, KRT16, and KRT107 [109, 110].

Ectodermal dysplasia is a separate one disorder similar to keratoderma or ichthyosis. It is associated with skin appendages bearing with the hair, nails, and sweat glands. Normally, this disorder is caused by a mutation in the KRT85 gene [111, 112].

2. Contribution of environmental constituents in mutation

Throughout the world, skin melanoma is the 19th most widespread cancer. Usually, almost all types of skin cancers are related to environmental factors including contact with immense ultraviolet radiation or due to sun exposure. Environmental mutagens may be synthetic or natural agents in nature [74, 113–116]. these mutagens produce genetic mutations or expand mutational activities during the life span [117, 118]. Most of the environmental mutagens possess genotoxic effects on the next generation via germ cells and continue in the inherent form.

Besides ultraviolet radiation, other radioactive, heavy metals, organic solvents or chemicals, viruses, bacteria, etc. also perform a role to cause cell damage [119–122]. Even, consumption of cigarette smoke, dietary contaminants including mycotoxin, aflatoxin B1, fat consumption, and unorganized stress are themselves integral environmental factors that contribute to cytokeratin disruptions [123]. All these agents are come in contact with the human through directly via skin & lungs or by ingestion. From this channel, these circulate in the body (blood, lymph glands, muscles, bones, tissues, and organs) and initiate mutations. In mutagenesis, these all mutagens penetrate directly to cellular and nuclear membranes and damage DNA by cross-linking (chemically gluing) two bases together. Sometimes, Mutagenesis is also caused by aberrant DNA methylation (epigenetic change) at the genomic level and post-translational modifications at the protein level. Finally, this results in genetic deficiencies cause.

3. Conclusion

Cytokeratins are essential protectors of epithelial structure found in the intracytoplasmic cytoskeleton of epithelial tissue. The occurrence of mutation in cytokeratin genes generates several genetic dermal disorders. In this genetic alteration, several environmental factors (ultraviolet, reactive oxygen species (ROS), deaminating agents, polycyclic aromatic hydrocarbon (PAH), ethylnitrosourea, azide, dyes, and heavy metals) take place as mutagen factors. The present chapter conclusively stated different cytokeratins disorders caused by environmental mutagens, i.e., synthetic or natural mutagens. In the end, based on environmental mutagens contributions, it can be stated that the genetic and epigenetic effects, arise/enhance through environmental mutagens/carcinogens, are the subject of innovative research.

Author details

Vishnu Sharma[1*], Tarun Kr. Kumawat[1], Garima Sharma[1], Rashi Garg[1] and Manish Biyani[1,2]

1 Department of Biotechnology, Biyani Girls College, Jaipur, Rajasthan, India

2 Department of Bioscience and Biotechnology, Japan Advanced Institute of Science and Technology, Japan

*Address all correspondence to: vishnusharma666@yahoo.com

References

[1] Chatterjee S. Cytokeratins in Health and Disease. 2012; 3(1): 198-202.

[2] Cooper D, Schermer A, Sun TT. Biology of disease. Classification of human epithelia and their neoplasms using monoclonal antibodies to keratins: Strategies, applications, and limitations. Laboratory Investigation. 1985; 52(3): 243-256.

[3] Herrmann H, Bär H, Kreplak L, Strelkov S V., Aebi U. Intermediate filaments: From cell architecture to nanomechanics. Nature Reviews Molecular Cell Biology. 2007; 8(7): 562-573.

[4] Franke WW, Schmid E, Osborn M, Weber K. Intermediate-sized filaments of human endothelial cells. Journal of Cell Biology. 1979; 81: 570-580.

[5] Oriolo AS, Wald FA, Ramsauer VP, Salas PJI. Intermediate filaments: A role in epithelial polarity. Experimental Cell Research. 2007; 313: 2255-2264.

[6] Schweizer J, Bowden PE, Coulombe PA, Langbein L, Lane EB, Magin TM, et al. New consensus nomenclature for mammalian keratins. Journal of Cell Biology. 2006; 174: 169-174.

[7] Weathersby C, Mcmichael A. Brazilian keratin hair treatment: A review. Journal of Cosmetic Dermatology. 2013; 12(2): 144-148

[8] Pan X, Hobbs RP, Coulombe PA. The expanding significance of keratin intermediate filaments in normal and diseased epithelia. Current Opinion in Cell Biology. 2013; 25:47e56.

[9] Wang F, Zieman A, Coulombe PA. Skin Keratins. In: Methods in Enzymology. 2016; 568: 303-350.

[10] Moll R, Divo M, Langbein L. The human keratins: biology and pathology. Histochemistry and cell biology. 2008; 129(6):705-33.

[11] Kumawat TK, Sharma A, Sharma V, Chandra S. Keratin Waste: The Biodegradable Polymers. In: Keratin. 2018.

[12] Bragulla HH, Homberger DG. Structure and functions of keratin proteins in simple, stratified, keratinized and cornified epithelia. In: Journal of Anatomy. 2009;214(4): 516-559.

[13] Vandebergh W, Bossuyt F. Radiation and functional diversification of alpha keratins during early vertebrate evolution. Molecular Biology and Evolution. 2012;29(3):995-1004

[14] Sharma V, Kumar Kumawat T, Seth R, Sharma A, Chandra S. A study on predominance of keratinophilic flora in soil of Jaipur, India. Journal of King Saud University - Science. 2020; 32(7): 2976-2981

[15] Heid HW, Moll I, Franke WW. Patterns of expression of trichocytic and epithelial cytokeratins in mammalian tissues II. Concomitant and mutually exclusive synthesis of trichocytic and epithelial cytokeratins in diverse human and bovine tissues (hair follicle, nail bed and matrix, lingual papilla, thymic reticulum). Differentiation. 1988; 37(3):215-30.

[16] Langbein L, Spring H, Rogers MA, Praetzel S, Schweizer J. Hair keratins and hair follicle-specific epithelial keratins. Methods in Cell Biology. 2004; 78:413-451.

[17] Langbein L, Schweizer J. Keratins of the human hair follicle. International Review of Cytology. 2005;243: 1-78.

[18] Schweizer J, Langbein L, Rogers MA, Winter H. Hair

follicle-specific keratins and their diseases. Experimental Cell Research. 2007; 313(10): 2010-2020.

[19] Alibardi L. Immunolocalization of alpha-keratins and feather beta-proteins in feather cells and comparison with the general process of cornification in the skin of mammals. Annals of Anatomy. 2013; 195(2): 189-198

[20] Calvaresi M, Eckhart L, Alibardi L. The molecular organization of the beta-sheet region in Corneous beta-proteins (beta-keratins) of sauropsids explains its stability and polymerization into filaments. Journal of Structural Biology. 2016; 194 (3): 282-291.

[21] Kreplak L, Aebi U, Herrmann H. Molecular mechanisms underlying the assembly of intermediate filaments. Experimental Cell Research. 2004; 301: 77 – 83.

[22] Rogers MA, Langbein L, Praetzel-Wunder S, Winter H, Schweizer J. Human Hair Keratin-Associated Proteins (KAPs). International Review of Cytology. 2006; 251:209-63.

[23] McKittrick J, Chen PY, Bodde SG, Yang W, Novitskaya EE, Meyers MA. The structure, functions, and mechanical properties of keratin. JOM. 2012; 64:449-468.

[24] Banerjee S, Wu Q, Yu P, Qi M, Li C. In silico analysis of all point mutations on the 2B domain of K5/K14 causing epidermolysis bullosa simplex: a genotype-phenotype correlation. Molecular Biosystems. 2014; 10(10):2567-2577.

[25] Yasukawa K, Sawamura D, McMillan JR, Nakamura H, Shimizu H. Dominant and recessive compound heterozygous mutations in epidermolysis bullosa simplex demonstrate the role of the stutter region in keratin intermediate filament assembly. Journal of Biological Chemistry. 2002; 277(26); 23670-23674.

[26] Chamcheu JC, Siddiqui IA, Syed DN, Adhami VM, Liovic M, Mukhtar H. Keratin gene mutations in disorders of human skin and its appendages. Archives of Biochemistry and Biophysics. 2011;508(2):123-137.

[27] Steinert PM. Structure, function, and dynamics of keratin intermediate filaments. Journal of Investigative Dermatology. 1993;100(6):729-734.

[28] Schweizer J, Bowden PE, Coulombe PA, Langbein L, Lane EB, Magin TM, et al. New consensus nomenclature for mammalian keratins. Journal of Cell Biology. 2006; 174(2): 169-174.

[29] Parry DAD. Hard α-keratin intermediate filaments: An alternative interpretation of the low-angle equatorial X-ray diffraction pattern, and the axial disposition of putative disulphide bonds in the intra-and inter-protofilamentous networks. International Journal of Biological Macromolecules. 1996; 19(1) 45-50.

[30] Rasool M, Nawaz S, Azhar A, Wajid M, Westermark P, Baig SM, et al. Autosomal recessive pure hair and nail ectodermal dysplasia linked to chromosome 12p11.1-q14.3 without KRTHB5 gene mutation. European Journal of Dermatology. 2010; 20(4): 443-446.

[31] Fraser RDB, MacRae TP, Suzuki E, Parry DAD, Trajstman AC, Lucas I. Intermediate filament structure: 2. Molecular interactions in the filament. Int J Biol Macromol. 1985; 7: 258-274.

[32] Herrmann H, Aebi U. Intermediate Filaments: Structure and Assembly. Cold Spring Harb Perspect Biol. 2016;8(11): a018242.

[33] Parry DAD, Fraser RDB. Intermediate filament structure: 1.

Analysis of IF protein sequence data. Int J Biol Macromol. 1985; 7: 203-213

[34] Bray DJ, Walsh TR, Noro MG, Notman R. Complete structure of an epithelial keratin dimer: Implications for intermediate filament assembly. PLoS ONE. 2015;10(7): e0132706.

[35] Deb-Choudhury S, Plowman JE, Harland DP. Isolation and Analysis of Keratins and Keratin-Associated Proteins from Hair and Wool. In: Methods in Enzymology. 2016; 568: 279-301.

[36] Fraser RDB, Parry DAD. Structural hierarchy of trichocyte keratin intermediate filaments. In: Advances in Experimental Medicine and Biology. 2018:57-70.

[37] Kim S, Coulombe PA. Intermediate filament scaffolds fulfill mechanical, organizational, and signaling functions in the cytoplasm. Genes and Development. 2007; 21: 1581-1597.

[38] Kornreich M, Avinery R, Malka-Gibor E, Laser-Azogui A, Beck R. Order and disorder in intermediate filament proteins. FEBS Letters. 2015; 589: 2464-2476.

[39] Eriksson JE, Dechat T, Grin B, Helfand B, Mendez M, Pallari HM, et al. Introducing intermediate filaments: From discovery to disease. Journal of Clinical Investigation. 2009; 119:1763-1771.

[40] Inagaki M, Inagaki N, Takahashi T, Takai Y. Phosphorylation-dependent control of structures of intermediate filaments: A novel approach using site- and phosphorylation state-specific antibodies. Journal of Biochemistry. 1997; 121: 407-414.

[41] Omary MB, Coulombe PA, McLean WHI. Intermediate Filament Proteins and Their Associated Diseases. N Engl J Med. 2004; 351:2087-2100.

[42] Omary MB, Ku NO, Liao J, Price D. Keratin modifications and solubility properties in epithelial cells and in vitro. Sub-cellular biochemistry. 1998.

[43] Kumar A, Jagannathan N. Cytokeratin: A reviewon current concepts. Int J OrofacBiol. 2018; 2: 6-11

[44] Steinert PM, Marekov LN, Fraser RDB, Parry DAD. Keratin intermediate filament structure. Crosslinking studies yield quantitative information on molecular dimensions and mechanism of assembly. Journal of Molecular Biology. 1993; 230(2): 436-452.

[45] Coulombe PA, Omary MB. "Hard" and "soft" principles defining the structure, function and regulation of keratin intermediate filaments. Current Opinion in Cell Biology. 2002; 14: 110-122.

[46] Jacob JT, Coulombe PA, Kwan R, Omary MB. Types I and II keratin intermediate filaments. Cold Spring Harbor Perspectives in Biology. 2018; 10(4): a018275.

[47] Chu PG, Weiss LM. Keratin expression in human tissues and neoplasms. Histopathology. 2002; 40(5): 403-439.

[48] Coulombe PA, Lee CH. Defining keratin protein function in skin epithelia: Epidermolysis bullosa simplex and its aftermath. Journal of Investigative Dermatology. 2012; 132: 763-775.

[49] Geisler N, Weber K. The amino acid sequence of chicken muscle desmin provides a common structural model for intermediate filament proteins. EMBO J. 1982; 1(12):1649-1656

[50] Lane EB, McLean WHI. Keratins and skin disorders. Journal of Pathology. 2004; 204(4):355-366.

[51] Parry DAD, Strelkov S V., Burkhard P, Aebi U, Herrmann H. Towards a molecular description of intermediate filament structure and assembly. Experimental Cell Research. 2007;313(10):2204-2216.

[52] Günther J, Seyfert HM. The first line of defence: insights into mechanisms and relevance of phagocytosis in epithelial cells. Seminars in Immunopathology. 2018; 40(6):555-565

[53] Slominski AT, Zmijewski MA, Skobowiat C, Zbytek B, Slominski RM, Steketee JD. Sensing the environment: Regulation of localand global homeostasis by the skin's neuroendocrine system. Advances in Anatomy Embryology and Cell Biology. 2012; 212 1-115

[54] Godsel LM, Hobbs RP, Green KJ. Intermediate filament assembly: dynamics to disease. Trends in Cell Biology. 2008; 18(1): 28-37.

[55] DeStefano GM, Christiano AM. The genetics of human skin disease. Cold Spring Harbor perspectives in medicine. 2014;4(10):a015172.

[56] Omary MB, Ku NO, Strnad P, Hanada S. Toward unraveling the complexity of simple epithelial keratins in human disease. Journal of Clinical Investigation. 2009; 119:1794-1805.

[57] Djudjaj S, Papasotiriou M, Bülow RD, Wagnerova A, Lindenmeyer MT, Cohen CD, et al. Keratins are novel markers of renal epithelial cell injury. Kidney International. 2016; 89(4): 792-808

[58] Rao RS, Patil S, Ganavi BS. Oral cytokeratins in health and disease. Journal of Contemporary Dental Practice. 2015; 15(1): 127-136

[59] Rugg EL, Leigh IM. The keratins and their disorders. American Journal of Medical Genetics Part C: Seminars in Medical Genetics. 2004; 131C(1): 4-11.

[60] Zhang L. Keratins in Skin Epidermal Development and Diseases. In: Keratin. 2018.

[61] Deo PN, Deshmukh R. Pathophysiology of keratinization. Journal of Oral and Maxillofacial Pathology. 2018; 22(1): 86-91.

[62] Shetty S, Gokul S. Keratinization and its disorders. Oman Medical Journal. 2012;27(5): 348-357.

[63] Bertolino AP, Checkla DM, Heitner S, Freedberg IM, Yu D-W. Differential Expression of Type I Hair Keratins. J Invest Dermatol. 1990; 94(3):297-303.

[64] Steinert PM. Keratins: Dynamic, flexible structural proteins of epithelial cells. Curr Probl Dermatol. 2001; 54:193-198

[65] Szeverenyi I, Cassidy AJ, Cheuk WC, Lee BTK, Common JEA, Ogg SC, et al. The human intermediate filament database: Comprehensive information on a gene family involved in many human diseases. Human Mutation. 2008; 29: 351-360.

[66] Hesse M, Zimek A, Weber K, Magin TM. Comprehensive analysis of keratin gene clusters in humans and rodents. Eur J Cell Biol. 2004; 83:19-26.

[67] Hesse M, Magin TM, Weber K. Genes for intermediate filament proteins and the draft sequence of the human genome: Novel keratin genes and a suprisingly high number of pseudogenes related to keratin genes 8 and 18. J Cell Sci. 2001; 114:2569-2575

[68] Pekny M, Lane EB. Intermediate filaments and stress. Experimental Cell Research. 2007; 313:2244-2254.

[69] Lane EB, Alexander CM. Use of keratin antibodies in tumor diagnosis. Seminars in cancer biology. 1990; 1(3):165-179.

[70] Moll R. Cytokeratins as markers of differentiation in the diagnosis of epithelial tumors. Sub-cellular biochemistry. 1998; 31:205-262.

[71] Nagle RB. A review of intermediate filament biology and their use in pathologic diagnosis. Molecular Biology Reports. 1994; 19(1):3-21.

[72] Oshima RG. Intermediate filaments: A historical perspective. Experimental Cell Research. 2007; 313(10):1981-1994.

[73] Schaafsma HE, Ramaekers FC. Cytokeratin subtyping in normal and neoplastic epithelium: basic principles and diagnostic applications. Pathology annual. 1994;29(Pt 1):21-62.

[74] Komine M. Regulation of Expression of Keratins and their Pathogenic Roles in Keratinopathies. In: Keratin. IntechOpen. 2018.

[75] Smith FJD, Kreuser-Genis IM, Jury CS, Wilson NJ, Terron-Kwiatowski A, Zamiri M. Novel and recurrent mutations in keratin 1 cause epidermolytic ichthyosis and palmoplantar keratoderma. Clinical and Experimental Dermatology. 2019;44(5): 528-534

[76] Zieman AG, Poll BG, Ma J, Coulombe PA. Altered keratinocyte differentiation is an early driver of keratin mutation-based palmoplantar keratoderma. Hum Mol Genet. 2019; 28(13

[77] Agarwala M, Salphale P, Peter D, Wilson NJ, Pulimood S, Schwartz ME, et al. Keratin 17 mutations in four families from India with pachyonychia congenita. Indian J Dermatol. 2017; 62(4): 422-426.

[78] Irvine AD, McKenna KE, Bingham A, Nevin NC, Hughes AE. A novel mutation in the helix termination peptide of keratin 5 causing epidermolysis bullosa simplex Dowling-Meara. J Invest Dermatol. 1997; 109: 815-816.

[79] Babu NA, Rajesh E, Krupaa J, Gnananandar G. Genodermatoses. Journal of Pharmacy and Bioallied Sciences. 2015; 7(Suppl 1): S203–S206.

[80] Arin MJ. The molecular basis of human keratin disorders. Human Genetics. 2009; 125:355-373.

[81] Smith FJD. The molecular genetics of keratin disorders. American Journal of Clinical Dermatology. 2003; 4(5):347-364.

[82] Uitto J, Richard G, McGrath JA. Diseases of epidermal keratins and their linker proteins. Experimental Cell Research. 2007; 313: 1995-2009.

[83] Bolling MC, Lemmink HH, Jansen GHL, Jonkman MF. Mutations in KRT5 and KRT14 cause epidermolysis bullosa simplex in 75% of the patients. Br J Dermatol. 2011; 164(3):637-644

[84] Fine J-D, Eady RAJ, Bauer EA, Bauer JW, Bruckner-Tuderman L, Heagerty A, et al. The classification of inherited epidermolysis bullosa (EB): Report of the Third International Consensus Meeting on Diagnosis and Classification of EB. Journal of the American Academy of Dermatology. 2008; 58(6): 931-950.

[85] Irwin McLean WH, Moore CB. Keratin disorders: From gene to therapy. Human Molecular Genetics. 2011; 20(R2): 189-197.

[86] Khani P, Ghazi F, Zekri A, Nasri F, Behrangi E, Aghdam AM, et al. Keratins and epidermolysis bullosa simplex. Journal of Cellular Physiology. 2018; 234(1):289-297.

[87] Natsuga K. Epidermolysis Bullosa Simplex. In Rijeka: IntechOpen; 2013. p. Ch. 2. Available from: https://doi.org/10.5772/54609

[88] Natsuga K, Nishie W, Shinkuma S, Arita K, Nakamura H, Ohyama M, et al. Plectin Deficiency Leads to Both Muscular Dystrophy and Pyloric Atresia in Epidermolysis Bullosa Simplex. Human Mutation. 2010; 31(10): E1687–E1698.

[89] Siañez-González C, Pezoa-Jares R, Salas-Alanis JC. Congenital epidermolysis bullosa: A review. ActasDermo-Sifiliograficas. 2009; 100(10):842-856

[90] Alper JC, Baden HP, Goldsmith LA. Kindler's syndrome. Arch Dermatol. 1978; 114(3): 457.

[91] Lai-Cheong JE, Tanaka A, Hawche G, Emanuel P, Maari C, Taskesen M, et al. Kindler syndrome: A focal adhesion genodermatosis. British Journal of Dermatology. 2009; 160(2):233-242.

[92] Souldi H, Bajja MY, Mahtar M. Kindler syndrome complicated by invasive squamous cell carcinoma of the palate. Eur Ann Otorhinolaryngol Head Neck Dis. 2018; 135(1):59-61.

[93] Youssefian L, Vahidnezhad H, Barzegar M, Li Q, Sotoudeh S, Yazdanfar A, et al. The Kindler syndrome: a spectrum of FERMT1 mutations in Iranian families. The Journal of investigative dermatology. 2015; 135: 1447-50.

[94] Hotz A, Oji V, Bourrat E, Jonca N, Mazereeuw-Hautier J, Betz RC, et al. Expanding the clinical and genetic spectrum of KRT1, KRT2 and KRT10 mutations in keratinopathic ichthyosis. ActaDermato-Venereologica. 2016; 96(4):473-478

[95] Ross R, DiGiovanna JJ, Capaldi L, Argenyi Z, Fleckman P, Robinson-Bostom L. Histopathologic characterization of epidermolytic hyperkeratosis: A systematic review of histology from the National Registry for Ichthyosis and Related Skin Disorders. J Am Acad Dermatol. 2008; 59: 86-90.

[96] Oji V, Tadini G, Akiyama M, Blanchet Bardon C, Bodemer C, Bourrat E, et al. Revised nomenclature and classification of inherited ichthyoses: Results of the First Ichthyosis Consensus Conference in Sorze 2009. Journal of the American Academy of Dermatology. 2010; 63(4) 607-641

[97] Dev T, Mahajan V, Sethuraman G. Hereditary palmoplantar keratoderma: A practical approach to the diagnosis. Indian Dermatology Online Journal. 2019; 10(4): 365

[98] Terron-Kwiatkowski A, Terrinoni A, Didona B, Melino G, Atherton DJ, Irvine AD, et al. Atypical epidermolytic palmoplantar keratoderma presentation associated with a mutation in the keratin 1 gene. British Journal of Dermatology. 2004; 150(6): 1096-1103

[99] Ferreira MAT, Gonzales JFO, Diniz BL, Floriani MA, Bau AEK, Rosa RCM, et al. Description of clinical aspects and microscopy of the hair shaft of a carrier of familial monilethrix. JornalBrasileiro de Patologia e Medicina Laboratorial. 2018; 54(5): 333-335.

[100] Van Steensel MA, Steijlen PM, Bladergroen RS, Vermeer M, van Geel M. A missense mutation in the type II hair keratin hHb3 is associated with monilethrix. Journal of medical genetics. 2005; 42(3): e19.

[101] Horev L, Glaser B, Metzker A, Ben-Amitai D, Vardy D, Zlotogorski A. Monilethrix: Mutational hotspot in the helix termination motif of the human hair basic keratin 6. Human Heredity. 2000; 50: 325-330

[102] Vikramkumar AG, Kuruvila S, Ganguly S. Monilethrix: A rare

hereditary condition. Indian J Dermatol. 2013; 58: 243

[103] Farooq M, Ito M, Naito M, Shimomura Y. A case of monilethrix caused by novel compound heterozygous mutations in the desmoglein 4 (DSG4) gene. British Journal of Dermatology. 2011; 165(2):425-431.

[104] Ferrando J, Galve J, Torres-Puente M, Santillán S, Nogués S, Grimalt R. Monilethrix: A new family with the novel mutation in KRT81 gene. International Journal of Trichology. 2012; 4(1): 53-55.

[105] Avhad G, Ghuge P, Jerajani H. Loose anagen hair syndrome. Indian Dermatology Online Journal. 2014;5(4):548-9.

[106] Agrawal D, Amin S, Adil M, Priya A. Uncombable hair syndrome with loose anagen syndrome: A rare association. Indian Journal of Paediatric Dermatology. 2020; 21(4):332-334.

[107] Srinivas SM. Loose Anagen Hair Syndrome. International journal of trichology. 2015;7(3):138-9.

[108] Ogunbiyi A. Pseudo folliculitisbarbae; current treatment options. Clinical, Cosmetic and Investigational Dermatology. 2019; 12: 241-247

[109] Shamsher MK, Navsaria HA, Stevens HP, Ratnavel RC, Purkis PE, Keisell DP, et al. Novel mutations in keratin 16 gene underly focal non-epidermolytic palmoplantar keratoderma (NEPPK) in two families. Human Molecular Genetics. 1995; 4:1875-1881.

[110] Smith FJD, Fisher MP, Healy E, Rees JL, Bonifas JM, Epstein EH, et al. Novel keratin 16 mutations and protein expression studies in pachyonychia congenita type 1 and focal palmoplantar keratoderma. Exp Dermatol. 2000; 9:170-177

[111] Prashanth S, Deshmukh S. Ectodermal Dysplasia: A Genetic Review. International Journal of Clinical Pediatric Dentistry. 2012; 5(3) 197-202

[112] Wright JT, Fete M, Schneider H, Zinser M, Koster MI, Clarke AJ, et al. Ectodermal dysplasias: Classification and organization by phenotype, genotype and molecular pathway. American Journal of Medical Genetics, Part A. 2019; 179(3):442-447

[113] Bernerd F, Del Bino S, Asselineau D. Regulation of keratin expression by ultraviolet radiation: Differential and specific effects of ultraviolet B and ultraviolet A exposure. Journal of Investigative Dermatology. 2001; 117(6): 1421-1429.

[114] Lopez A, Liu L, Geskin L. Molecular Mechanisms and Biomarkers of Skin Photocarcinogenesis. In 2018.

[115] Narayanan DL, Saladi RN, Fox JL. Ultraviolet radiation and skin cancer. International Journal of Dermatology. 2010;49: 978-986.

[116] Saladi RN, Persaud AN. The causes of skin cancer: A comprehensive review. Drugs of Today. 2005; 41(1): 37-53.

[117] Ralston A. Environmental mutagens, cell signalling and DNA repair. Nature Education. 2008; 1(1): 114.

[118] Yagi T. A perspective of Genes and Environment for the development of environmental mutagen research in Asia. Genes and Environment. 2017; 39: 23.

[119] Sesto A, Navarro M, Burslem F, Jorcano JL. Analysis of the ultraviolet B response in primary human keratinocytes using oligonucleotide microarrays. Proceedings of the

National Academy of Sciences of the United States of America. 2002; 99(5): 2965-2970.

[120] D'Orazio J, Jarrett S, Amaro-Ortiz A, Scott T. UV radiation and the skin. International Journal of Molecular Sciences. 2013; 14:12222-12248.

[121] Chatterjee N, Walker GC. Mechanisms of DNA damage, repair, and mutagenesis. Environmental and Molecular Mutagenesis. 2017; 58: 235-263.

[122] Hartwig A, Arand M, Epe B, Guth S, Jahnke G, Lampen A, et al. Mode of action-based risk assessment of genotoxic carcinogens. Archives of Toxicology. 2020; 94: 1787-1877.

[123] Minamoto T, Mai M, Ronai Z. Environmental factors as regulators and effectors of multistep carcinogenesis. Carcinogenesis. 1999 Apr;20(4):519-27.